Urban Sustainability

The Urban Sustainability Book Series is a valuable resource for sustainability and urban-related education and research. It offers an inter-disciplinary platform covering all four areas of practice, policy, education, research, and their nexus. The publications in this series are related to critical areas of sustainability, urban studies, planning, and urban geography.

This book series aims to put together cutting-edge research findings linked to the overarching field of urban sustainability. The scope and nature of the topic are broad and interdisciplinary and bring together various associated disciplines from sustainable development, environmental sciences, urbanism, etc. With many advanced research findings in the field, there is a need to put together various discussions and contributions on specific sustainability fields, covering a good range of topics on sustainable development, sustainable urbanism, and urban sustainability. Despite the broad range of issues, we note the importance of practical and policy-oriented directions, extending the literature and directions and pathways towards achieving urban sustainability.

The series will appeal to urbanists, geographers, planners, engineers, architects, governmental authorities, policymakers, researchers of all levels, and to all of those interested in a wide-ranging overview of urban sustainability and its associated fields. The series includes monographs and edited volumes, covering a range of topics under the urban sustainability topic, which can also be used for teaching materials.

Saeid Pourroostaei Ardakani · Ali Cheshmehzangi

Big Data Analytics for Smart Urban Systems

Springer

Saeid Pourroostaei Ardakani
School of Computer Science
University of Lincoln
Lincoln, UK

Ali Cheshmehzangi
Department of Architecture
Qingdao City University
Qingdao, China

ISSN 2731-6483 ISSN 2731-6491 (electronic)
Urban Sustainability
ISBN 978-981-99-5542-8 ISBN 978-981-99-5543-5 (eBook)
https://doi.org/10.1007/978-981-99-5543-5

This Springer imprint is published by the registered company Springer Nature Singapore Pte Ltd.
The registered company address is: 152 Beach Road, #21-01/04 Gateway East, Singapore 189721, Singapore

Paper in this product is recyclable.

We collectively dedicate this book to resilient societies and economies that suffer from various hardships, inhumane situations, challenging living environments, and devastating circumstances. Millenniums of cultural and societal development should not be sabotaged by just a few decades of decay, decline, and corrosion. We trust there is light at the end.

Preface

Data is the new science. Big Data holds the answers. Are you asking the right questions?

—Patrick P. Gelsinger

In the age where data dictates many things, we must pay attention to the critical role of data science in managing and governing our contemporary society. Whether we like it or not, data science has become more progressive in recent one or two decades, penetrating various sectors and through the use of multiple methods. While contextual and cultural factors have played their parts hand-in-hand with how data is collected, analysed, and used, we note there are still some common or standardised grounds between data science and its (increasing) impact on global research and practice. One of these common grounds is the diverse range of effective methods used in big data analytics, particularly in optimising, managing, and helping decision-making processes.

The nexus between big data analytics and smart urban systems interests many stakeholders. For researchers and practitioners, this era creates a unique opportunity for interdisciplinary, cross-disciplinary, and even trans-disciplinary studies. Integration remains key, and data remains the source for making the integration happen, and if so, how and in which direction it could make a difference to society. Thus, in this book, we combine computer science and urbanism knowledge to better evaluate and understand urban systems and their operations. In particular, we highlight 'smartness' and 'smart urban systems' to consider improving cities, city processes, and operations. In such revolutionary transformations that we go through, innovation has played a vital part in making data available and utilised in new realms of cross-disciplinary research.

Cities and communities are exciting living networks where data exchanges occur constantly, data is fluid and collected simultaneously, and data is shared and re-shared with or without our consent. Knowledge and data continuously pour into multiple mediums of data collection centres, various personal and public devices, sensor networks, etc. While we are aware of these interminable transactions, we must recognise the value behind how data analytics could benefit it in the longer term. Set the cons aside, we could see more progress in this area, particularly in the smart city movement, where smartness, smart communities, and smart everything floats around our daily lives. Hence, we aim to provide a range of case study examples focused on societal and economic aspects of cities, where big data analytics and smart urban systems are closely correlated and optimisation, decision-making, and management processes are tangible.

Big data analytics for smart urban systems is the first volume of our two book projects exploring global case study examples in big data research. As the title entails, we focus on methodological interventions and contributions that inspire future research and practice directions. This volume comprehensively covers big data solutions in smart urban systems and applications, dealing with data analytics problems, methods, and approaches relevant to research and practice. It also describes machine learning solutions to handle large and rapidly changing data in cities. These are the backbone of the future big data analytics research and practice should we decide to continue with advancing smartness, smart development, and smart urban systems.

Lincoln, UK
Qingdao, China
June 2023

Saeid Pourroostaei Ardakani
Ali Cheshmehzangi

Acknowledgements

We collectively acknowledge our research interns, team members, and external collaborators. All members have been extremely helpful in completing this book project. Despite the hardships, they have provided us with excellent support in collecting data, conducting research, and completing the tasks under each given project. Our research interns worked hard according to the assigned tasks and objectives. Our team members and external collaborators helped with reviews and gave us valuable feedback. We thank them all for their continuous support and hope to have longer-term relationships with them.

We reflect on these two incredibly fruitful years by confirming that our choices were made to ensure that the right paths were selected.

Ali Cheshmehzangi acknowledges the Ministry of Education, Culture, Sports, Science, and Technology (MEXT), the Japanese Government, and Hiroshima University, Japan.

About This Book

Big Data Analytics for Smart Urban Systems aims to introduce Big data solutions for urban sustainability smart applications, particularly for smart urban systems. It focuses on intelligent big data which takes the benefits of machine learning to analyse large and rapidly changing datasets in smart urban systems. The state-of-the-art Big data analytics applications are presented and discussed to highlight the feasibility of big data and machine learning solutions to enhance smart urban systems, smart operations, urban management, and urban governance. The key benefits of this book are, (1) to introduce the principles of machine learning-enabled big data analysis in smart urban systems, (2) to present the state-of-the-art data analysis solutions in smart management and operations, and (3) to understand the principles of big data analytics for smart cities and communities.

Praise for *Big Data Analytics for Smart Urban Systems*

"Over the many years of collaboration between academia and industry, we noticed the common language is 'big data'; with that, we have developed novel ideas to bridge the gaps and help promote innovation, technologies, and science".

—Tian Tang, *Independent Researcher, China*

"Big Data Analytics is a fascinating research area, particularly for cities and city transformations. This book is valuable to those who think vigorously and aim to act ahead".

—Li Xie, *Independent Researcher, China*

"For urban critiques, knowledge trains aspiring opportunities toward outstanding manifestations. Smartness has evolved or/advanced rambunctious and embracing realities along (with) novel directions and nurturing integrated city knowledge".

—Aaron Golden, *SELECT Consultants, UK*

Contents

About the Authors

Saeid Pourroostaei Ardakani currently works as a senior lecturer of Computer Science at the University of Lincoln, UK. He is also an associated academic member of Lincoln Centre for Autonomous Systems (L-CAS) and has formerly worked at the University of Nottingham (UNNC) and Allameh Tabatabai University (ATU), as an assistant professor of Computer Science, a member of the Next Generation Internet of Everything Laboratory (NGIoE) and Artificial Intelligent Optimisation Research group, and the head of ATU-ICT centre. He received his Ph.D. in Computer Science from the University of Bath focusing on data aggregation routing in Wireless Sensor Networks. Saeid's research and teaching expertise centres on smart and adaptive computing and/or communication solutions to build collaborative/federated (sensory/feedback) systems in Internet of Things (IoT) applications and cloud environments. He is also interested in (ML-enabled) big data processing and analysis applications. Saeid has published more than 60 scholarly articles in reputed international journals and peer-reviewed conferences.

Ali Cheshmehzangi is the world's top 2% field leader, recognised by Stanford University. He has recently taken a senior leadership and management role at Qingdao City University (QCU), where he is a professor in Urban Planning, the director of the Center for Innovation in Teaching, Learning, and Research, and the advisor to the school's international communications. Over 11 years at his previous institute, Ali was a full professor in Architecture and Urban Design, the head of the Department of Architecture and Built Environment, the founding director of the Urban Innovation Lab, the director of Center for Sustainable Energy Technologies, and the director of Digital Design Lab. He was the visiting professor and now a researchaAssociate of the Network for Education and Research on Peace and Sustainability (NERPS) at Hiroshima University, Japan. Ali is globally known for his research on 'urban sustainability'. So far, Ali has published over 300 journal papers, articles, conference papers, book chapters, and reports. To date, he has 15 other published books.

Chapter 1
Big Data Analytics: An Introduction to Their Applications for Smart Urban Systems

1.1 The Emergence of Big Data Analytics

Over the last two decades, we have witnessed how analytics and data science have changed how we think about our urban systems. With the rise of smart cities and smart urban systems, in particular, we see the emergence of big data analytics to be more progressive and towards changes in management, governance, and culture of urban thinking. Although big data is usually unstructured, data science has been the foundation of new ways to acquire, utilize, and analyze data. In these two decades, big data analytics have become more important, leading toward new analysis methods, optimisation, management, and productivity. Thus, the emergence of big data analysis has already passed its golden time, and the shift has become more aligned with integrated solutions, ICT-driven or ICT-based approaches [6, 7], and technology-oriented methods [18]. As Davenport [11] puts it well, the age of analytics technology has already changed the way we work analytically:

> Another key change in the analytics technology landscape involves autonomous analytics—a form of artificial intelligence or cognitive technology. Analytics in the past were created for human decision makers, who considered the output and made the final decision. But machine learning technologies can take the next step and actually make the decision or adopt the recommended action. Most cognitive technologies are statistics-based at their core, and they can dramatically improve the productivity and effectiveness of data analysis.

In this regard, the evolution of analytics is progressive and integrated into all systems, including smart urban systems. Thus, it is inevitable to see some major paradigm shifts, particularly those based on strategical management changes, potential technology integration, and an entirely new epistemological approach for data-driven decision-making and re-orientation processes.

In general, we note that cities have always been playful grounds to test out various ways of big data analysis, data science integration, and analytical technologies. The shift towards resilient and smart cities [8, 15, 22] has become ever-important, particularly since urban analytical approaches/methods have become more viral in contemporary (urban and non-urban) research and practices. From social media big data

S. Pourroostaei Ardakani and A. Cheshmehzangi, *Big Data Analytics for Smart Urban Systems*, Urban Sustainability, https://doi.org/10.1007/978-981-99-5543-5_1

analytics to decision-making processes for cities and regions, big data has become the backbone of optimisation, management, and governance. In recent years, urban officials and experts have become more open to considering big data analytics for strategic planning, design interventions, and even policy changes.

Associated with this overarching emergence of big data analytics, we bring together a range of big data analytics examples to explore and discuss how cities and urban systems could become optimised, and, ultimately, managed in better ways. There are, of course, pros and cons of such methods, but we note there are more benefits than harms, those that could constructively help us bring data science and urban systems closer to each other. The already progressive examples of algorithmic urban planning and various methods of big data use in urban development suggest how we are already transitioning to smart urban management and governance. We have to consider such transition, or eventual transformations, with a pinch of salt, knowing that recent and ongoing changes may significantly impact how cities are developed, managed, and run. Such changes may not necessarily be pleasant, but some will be welcoming and forward-thinking. However, we are still waiting for longer-term impacts due to such changes or transformations.

This chapter serves as the introduction chapter of this book, allowing us to highlight a few essential points regarding the aim and objectives of the book, the book structure (as well as the introduction of two volumes under the big data analytics research), and finally, a summary of the following case study chapters.

1.2 The Aim and Objectives of the Book

This book aims to put together a comprehensive set of case study examples in various categories related to smart urban systems, smart operations, smart governance, and smart urban management. All these examples highlight how we now use urban data analytics through research and practice [3], which is highly important to more significant movements such as smart cities, smart development, and smartness. Thus, the book aims to introduce main methods rather than particular cases. However, it uses global case study examples to demonstrate how these methods are utilised for various (smart) urban systems. In doing so, the ultimate objectives or goals are to introduce the principles of machine learning-enabled big data analysis in smart urban systems, to present state-of-the-art data analysis solutions in smart management and operations, and to understand the principles of big data analytics for smart cities and communities (Fig. 1.1). By demonstrating various methods, we would shed light on how big data analytics play an influential part in achieving, developing, or promoting smart urban systems.

The importance of data and how it can be used extensively show new directions in big data research, particularly beyond just big data analytics. We explore examples that suggest data exploration, data preparation, data modelling, and data evaluation. All these aspects are vital to how big data moves urban-related research forward. From the small smart mobile devices we carry daily to large monitoring and data

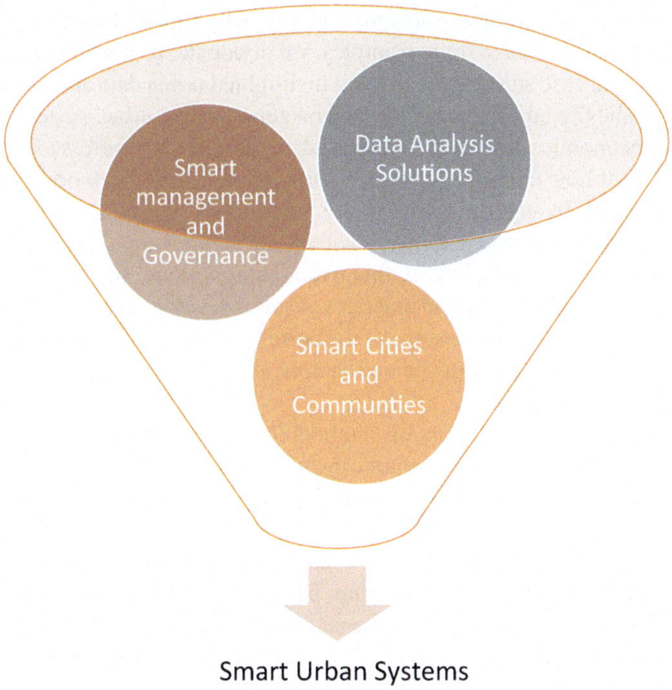

Fig. 1.1 The state-of-the-art data analysis solutions in smart management and operations, and to understand the principles of big data analytics for smart cities and communities

collection centers, data is continuing to be collected and explored. There are many ways data collection is used or become functional for various purposes, often for control and management, but also to optimise operations and enhance how everyday life passes by. In recent years, the data analytics funnel has shifted to focus more on decreasing content volume while increasing content quality, meaning big data is now more than just public data, sensor data, structured data, and unstructured data. In structuring and linking data across various sectors, we are able to move toward integrating data science with various purposes and functionalities, i.e., to optimise urban systems and help shape better urban management and governance.

1.3 The Structure of Two Volumes on Big Data Analytics

Aligned with the aim and objectives of the book, we focus on critical areas of research and practice in smart urban systems. These areas are divided into two separate—but correlated—volumes, covering the big data analytics topic. At first, we wanted to focus on urban sustainability aspects and direction, but we realised the work fits well with smart urban systems, particularly that data science is closely related to smart practices or smartness in cities and communities. To comprehensively cover various methods and smart urban system examples, we divide the project into two volumes, each with several case study examples and highlighted urban data analytics methods. The first volume (i.e., this book) is focused on genera smart urban systems, precisely focused on economic and societal case study examples. We note such aspects are the backbone of how cities are sustained and remain liveable in working and living environments [5]. In suggesting the future smart cities, we note the importance of economic development and stability, societal development, and healthy operations of urban systems. In light of these, this overarching volume focuses on case study examples related to the theory and practices of smart development and urban systems. In this volume, we also explore some of the correlated factors to urban society issues, which refers to a diverse set of societal factors and opportunities for societal smart city and societal sustainability.

In the first volume, we focus on methodological contributions to smart urban systems. Beyond just developing smart urban networks, we hope to see further development in data-driven urban development opportunities that benefit from optimising or enhancing urban systems [2, 16]. In various reports on 'building a smart city', we see examples of how technologies have been utilised to make our cities and communities smarter. Yet, we see more than that, particularly aligned with data-based and data-driven directions. Hence, this volume explores examples of big data analytics through their methodological contributions in making urban systems smarter and more resilient.

By going back to the original smart planning framework ideology, we note the importance of systems thinking, collaborative methods, ubiquitous digitalisation, and knowledge management in conducting big data analytics studies in smart urban systems. In recent research, we see a growing trend of big data analytics in smart urban metabolism studies [17], smart city applications [4, 20], and specific sectors in cities such as transportation [21]. Hence, we see a growing trend in ICT-based or ICT-driven applications, including digital data acquisition, collection, analysis, and usage [10]. In this volume, we aim to adopt a various range of data-driven methods [9] for different global case study examples. In doing so, we mainly focus on methodological contributions rather than sectoral contributions. However, we focus on key aspects or areas associated with smart urban systems with the aim to optimise smart management and governance of cities and communities.

In the cycle of big data for smart cities, we note the importance of having access to valid data at the first stage, then being able to analyse and make decisions based on the analysis, and finally collecting data from various means and opportunities (e.g., sensors, machines and equipment, apps, public data, devices, etc.) (See Fig. 1.2). According to Aymen et al. [1], a smart city is relatively complex and is a "*living space requiring technologies such as high-tech information to improve the quality of residents' lifestyle and to optimize, economically and environmentally, the management of available resources such as roads, building activities, environments, lights and water*". In their study (ibid), they show the importance of various data collection sources, such as wireless sensor networks, smart city monitoring systems, use of technologies and participants, the Internet of Things (IoT), etc.

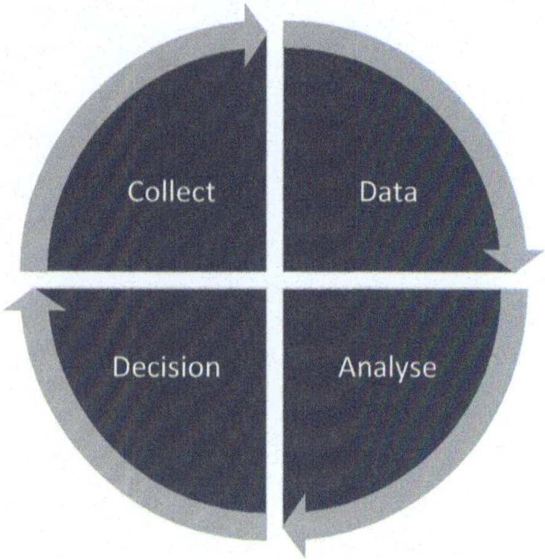

Fig. 1.2 Big Data and smart city (Redrawn by the authors and extracted from the conceptual demonstration by Aymen et al. [1])

Based on the findings from the literature, we see a robust linkage between data science, data discovery, and big data (see Fig. 1.3). They are also differences between the three, showing that "*each of these areas has seen explosive growth, but there are clear upsides and downsides to each. For example, Data Discovery excels in ease of use, but allows only limited depth of exploration, while Data Science provides powerful analysis but is slow, complex, and difficult to implement*" [12]. Hence, the overarching areas of data mining and knowledge discovery have developed further into big data analytics and knowledge extraction [19], commonly more advanced in four disciplines of mathematics, statistics, computer science, and computational science. More recently, such approaches have excelled in other fields, particularly in urban studies, urban analytics, and urban planning/design.

Below, based on the suggestions from EPM Channel [12], we summarise the pros (Fig. 1.4) and cons (Fig. 1.5) of each of these main areas, i.e., data science, data discovery, and big data.

In the second volume, we focus more on specific sectors beyond just the general urban systems. In the following book, we delve into two areas of 'smart transport' and 'smart healthcare'. For smart transport systems, we explore various big data analytics methods and studies related to transportation systems and networks, which are extremely important in how cities operate. Mobility, movability, accessibility, and connectivity within and between cities and regions are highly important in various ways. For many years, big data for transportation [13] has been used for smart mobility and management, including ways and methods of developing future smart transportation systems, smart networks, and integrated solutions for transportation sustainability. In the other part of the second volume, we highlight case study examples related to (smart) healthcare, a significant part of how cities ought to become healthier for all, allowing us to explore more than just smart health communities, digital health, and high-tech urban interventions.

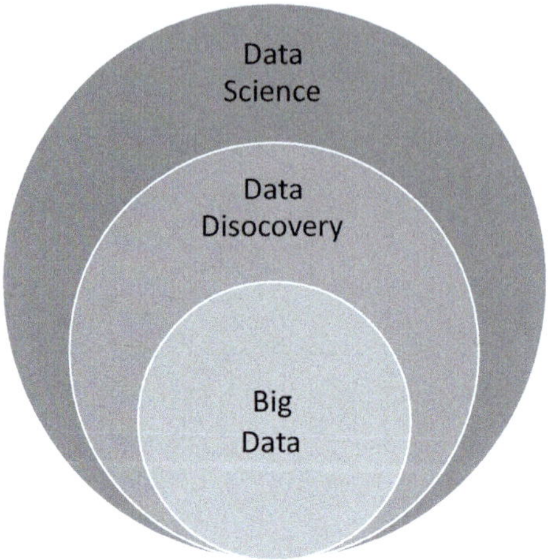

Fig. 1.3 The correlation between data science, data discovery, and big data, inspired from the study covered by EPM Channel indicating that big data discovery is the next trend in data analytics research and practice. This is particularly important to companies and businesses where they require to use predictive analytics, promote easy-to-use data preparation and mining, and consider the combination of traditional advanced analytics

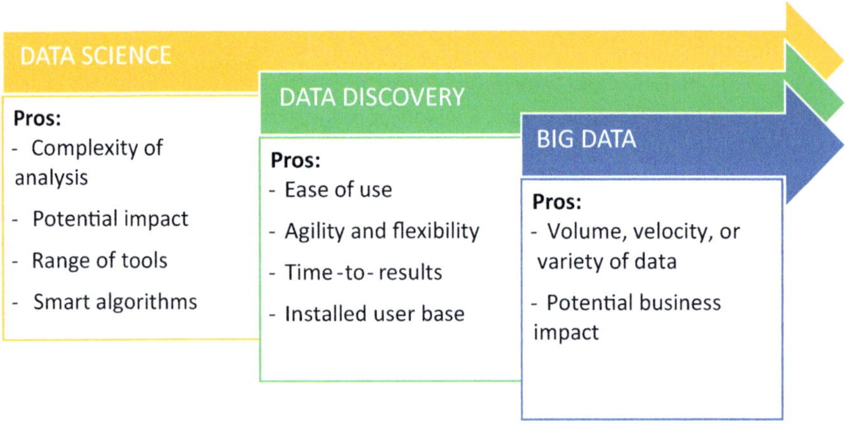

Fig. 1.4 Pros of three areas of 'data science', 'data discovery', and 'big data' (Redrawn by the authors from the original discussions by EPM Channel [12])

1.4 A Summary

This book contributes to theories and methods of big data analytics in cities and smart urban systems. In the following chapter, we explore case study examples and various big data analytics methods. These methods are highlighted as major methods in the field, with great potential or scope for further development and integration in smart

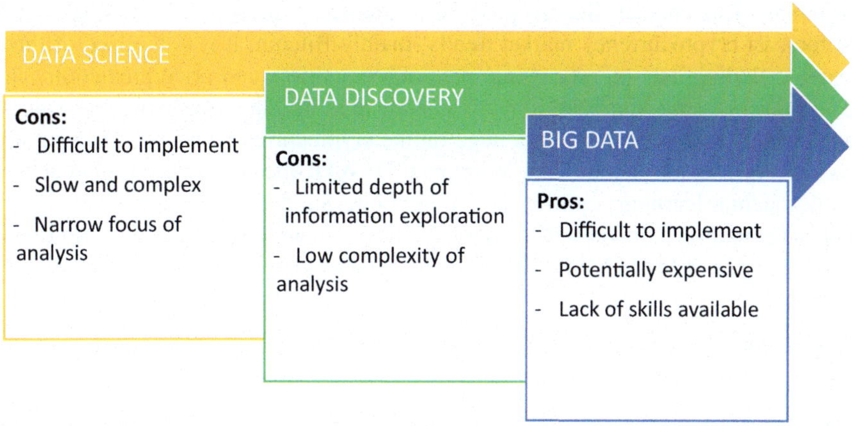

Fig. 1.5 Cons of three areas of 'data science', 'data discovery', and 'big data' (Redrawn by the authors from the original discussions by EPM Channel [12])

urban management and governance scenarios. While we mainly focus on methodological contributions, global case study examples are used to present, visualise, and summarise big data analytics methods in various urban systems, conditions, sectors, and contexts. The following six book chapters are summarised below.

Chapter 2: Stock Market Prediction During COVID-19 Pandemic: A Time-Series Big Data Analysis Method

Abstract

Stock price prediction is one of the most difficult fields to study because of the irregularities. With the outbreak of COVID-19, the epidemic has also greatly affected the stock market. Because stock prices sometimes show similar patterns and are determined by a variety of factors, we propose identifying comparable patterns in past stock data of daily stock prices, as well as selecting the primary components that significantly affect the price, including the impact of COVID-19. In this research, we focus on using big data methods to analyse the patterns of stock price and use the LSTM model for stock prediction and analysed the impact of the epidemic on stock trends.

Keywords: Stock market prediction; Min–Max Normalisation; Time-series data analysis; LSTM; COVID-19.

Chapter 3: A Big Data Solution to Predict Cryptocurrency Market Trends: A Time-series Machine Learning Approach

Abstract

Cryptocurrency trend analysis allows researchers to study cryptocurrency market behaviour and propose predictive statistical, machine learning, and/or economic solutions. This chapter aims to propose a time-series machine learning approach to forecast cryptocurrency market trends -mainly Bitcoin. It is comprised of three steps, including data preprocessing, pattern recognition, and price prediction. The data preprocessing approach aims to handle missing data, normalise the values, and convert time-series data in a huge online cryptocurrency dataset. Data pattern recognition uses a Dynamic Time Wrapping (DTW) K-means clustering approach to, while machine learning perdition models use Long-Short Term Memory and Random Forest techniques to forecast Bitcoin price. The predictive models are tested and evaluated in terms of MAE, MSE, RMSE and R 2 scores to find the best-fitted prediction approach.

Keywords: Cryptocurrency Market; K-means; Time-series Pattern recognition; Dynamic Time Wrapping; Random forest.

Chapter 4: Big Data Analytics for Credit Risk Prediction: Machine Learning Techniques and Data Processing Approaches

Abstract

Credit risk scoring approaches study and analyse customer's financial records to provide the financial institutions with a summarised decision making information. However, they still suffer from the lack of a solid Big Data solutions to recognise, model and predict credit risk data patterns. This chapter aims to propose machine learning pipelines which are capable of extracting principal information from a huge and public credit risk dataset. For this, a Big data-enabled data preprocessing approach is proposed to prepare the given dataset. Moreover, two machine learning models including Decision Tree and Gradient Boosting are trained, tested and evaluated to find the best-fitted techniques for credit risk prediction. According to the results, Gradient Boosting (AUC of 0.987) gives a better performance as compared to Decision Tree (AUC of 0.488).

Keywords: Decision Tree; Gradient Boosting; AUC; Mortgage; Credit Risk Scoring.

Chapter 5: Worldwide Mobility Trends and the COVID-19 Pandemic: A Federated Regression Analysis During the Pandemic's Early Stage

Abstract

The COVID-19 pandemic has changed the world, and people have experienced various restrictions and situations since January 2020. This chapter highlights the impact of the pandemic on people's mobility trends at the early stages. It uses a correlation matrix method to find the correlations of mobility trends and six commonly-used places, including retail and recreation, groceries and pharmacies, parks, transit stations, workplaces, and residential once the COVID-19 outbreak. It proposes a regression-enabled machine learning model to predict daily travel using the current epidemic situation. This model offers the governmental, industrial, and business sectors benefits to propose a plan to manage resources and provide services with minimised outbreaks in the future.

Keywords: COVID-19; Mobility trends; Big data; Regression; Correlation matrix.

Chapter 6: Adaptive Feature Selection for Google App Rating in Smart Urban Management: A Big Data Analysis Approach

Abstract

Most feature selection algorithms use a similarity matrix to assign a fixed value to pairs of objects. However, if the dataset is large and includes noisy or unlabeled samples, they could produce incorrect or inaccurate conclusions. This chapter refers to a single target prediction analysis on Google's App rating. This aspect is relevant to smart urban management and systems and could help optimize data analysis for multiple uses. After conducting necessary experiments, the findings from this chapter show that the adaptive feature selection (e.g., Random Forest) gives optimized results compared to traditional feature selection techniques, such as 'Linear Discriminant Analysis' and 'Principal Component Analysis'. The results from this comparison help future research in big data analysis, particularly for urban management and smart urban systems.

Keywords: Feature selection; Random forest; Principal component Analysis; Linear discriminant analysis; Machine learning.

Chapter 7: Improve the Daily Societal Operations using Credit Fraud Detection: A Big Data Classification Solution

Abstract

Credit fraud detection grows rapidly with the development of electronic payment. It is part of the smart management in daily societal operations. IEEE-CIS Fraud Detection dataset was raised for technique innovation to tackle this problem. For the consideration of the size of the available fraud dataset, this chapter proposes a solution with the combination of the big data technique and the machine learning algorithms. Aside from the raised pipeline of the solution, other identified works include a comparison of four implemented machine learning algorithms, three surveys into the computing time with respect to the data size, number of executors, and number of cores. Findings from this chapter help finding solutions for more efficient credit fraud detection.

Keywords: Index Terms; Credit fraud detection; Machine learning; Spark MLlib; PCA.

Lastly, the book concludes with a summary of lessons learnt and future research directions, indicating the role of big data analytics in optimising smart urban systems. By concluding the book, we shed light on the book's methodological contributions as well as how the findings could help direct future urban-related research in correlation with data science and big data analytics. At the end of this chapter, we provide a set of boxes to highlight some of the recent examples of smart city documents and reports across multiple sectors and from various stakeholders in the industry, government, etc.

Box 1.1 Examples of 'Smart Cities' reports and documents

2017 report

The Smart Cities special report published in The Times—Raconteur

"The phrase "smart city" turns 25 this year, but what does "smart" really mean for cities? The Smart Cities special report, published in The Times, unpicks what the future city will look like, as well as the current market for smart solutions. The report covers the latest technology helping to solve urban issues, Latin America's first smart city and how personal data can be used to improve citizens' lives. In addition, it includes a pull-out annotated blueprint of the Shanghai Tower, one of the world's most cutting-edge sustainable buildings".

Available from: https://www.raconteur.net/smart-cities-2017.

Box 1.2 Examples of 'Smart Cities' reports and documents

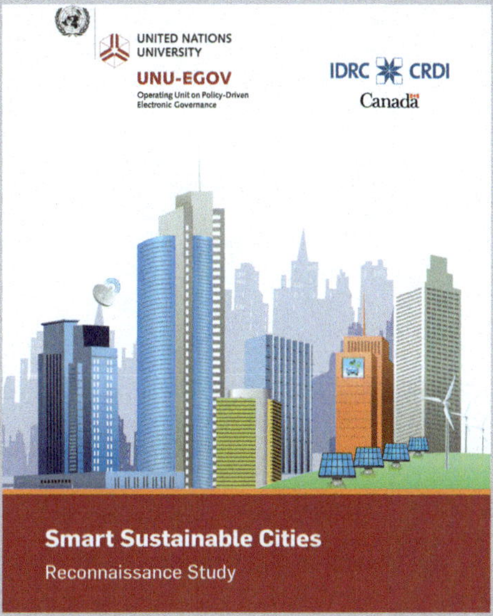

Smart Sustainable Cities—Reconnaissance Study

"Smart Cities have emerged as one response to the challenges and opportunities created by rapid urbanization. The report "Smart Sustainable Cities—Reconnaissance Study" presents the results of a study, conducted by UNU-EGOV and funded by IDRC, that examined the thesis that Smart Cities advance sustainable development. The study analysed 876 scientific publications, recommendations from 51 think tank organizations and 119 concrete Smart City initiatives. Researchers also conducted several interviews with city managers, planners and researchers responsible for successful Smart City initiatives. The full report is available at UNU Collections".

Available from: https://egov.unu.edu/news/smart-sustainable-cities-reconaissance-study.html.

Box 1.3 Examples of 'Smart Cities' reports and documents

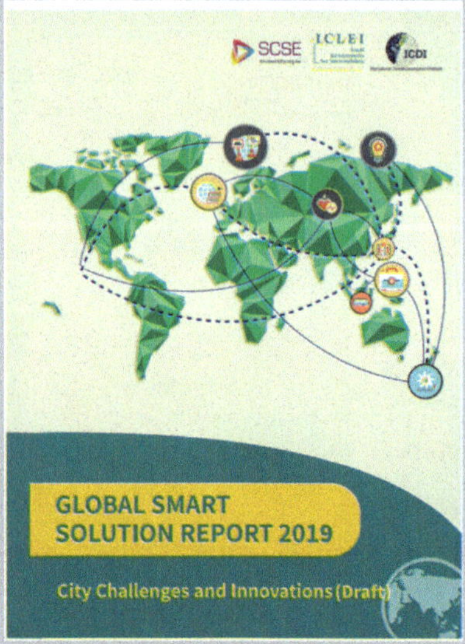

2019 report

Global Smart City Report Released in COP24

"Taiwan Smart City Solutions Alliance (TSSA), International Climate Development Institute (ICDI), and ICLEI Kaohsiung Capacity Center (ICLEI KCC) worked together to release the Global Smart Solution Report 2019 (GSSR 2019) in the side event during the UNFCCC/COP24 meeting in Katowice, Poland. ICDI Executive Director Kung-Yueh Camyale Chao said, there were 20 cities all over the world to submit the application. After reviewed by 7 experts, there are 12 programs from 8 cities have been included in the first-year report. The standards to choose the cases are Economic Development, Social Justice, and Ecological Responsibility. In addition, the programs need to identify the problem of the city and provide smart solutions which can also respond to UN Sustainable Development Goals (SDGs)".

Available from: https://en.smartcity.org.tw/index.php/en-us/posts/news/item/70-global-smart-city-report-released-in-cop24.

Box 1.4 Examples of 'Smart Cities' reports and documents

Growing Smart Cities in Denmark

"Over the last decade the 'smart city' concept has emerged to represent technology-driven urban benefits and the products and services that deliver them. For national governments, the smart city is attractive because it represents an opportunity to improve its towns and cities and to access a large global market, estimated to be in the order of $1.3 trillion and growing by 17% each year.

National governments are ramping up their efforts to remove barriers that are preventing regional and municipal governments from applying smart city solutions and local businesses from developing and exporting related products and services. This paper explores smart cities projects that have taken place in Denmark looking at four very different Danish cities (Copenhagen, Aarhus, Vejle, Albertslund) shows the widespread pursuit of smart city benefits by all of these players and identifies some of the obstacles to adoption that are faced".

Available from: https://www.arup.com/perspectives/publications/research/section/growing-smart-cities-in-denmark.

Box 1.5 Examples of 'Smart Cities' reports and documents

Smart City Strategies: A Global Review

"The pervasive nature of digital technology means that cities are constantly evolving to meet the changing needs of everyday life. The term 'smart city' was adopted in the early 2010s to describe the increasing use of technology and data to inform decision making for governing cities. This report, produced by Future Cities Catapult and Arup, explores the landscape of smart city strategies and aims to provide insight into how cities around the world are approaching the smart city agenda. The review enables us to build a richer knowledge base for cities and shed light on the principles and patterns seen in smart city strategies across the globe. Twenty-one cities of varying geography, population and stage of existing smart city strategy were studied, including New York, Berlin, São Paulo and Manchester".

Available from: https://www.arup.com/perspectives/publications/research/section/smart-city-strategies-a-global-review.

Box 1.6 Examples of 'Smart Cities' reports and documents

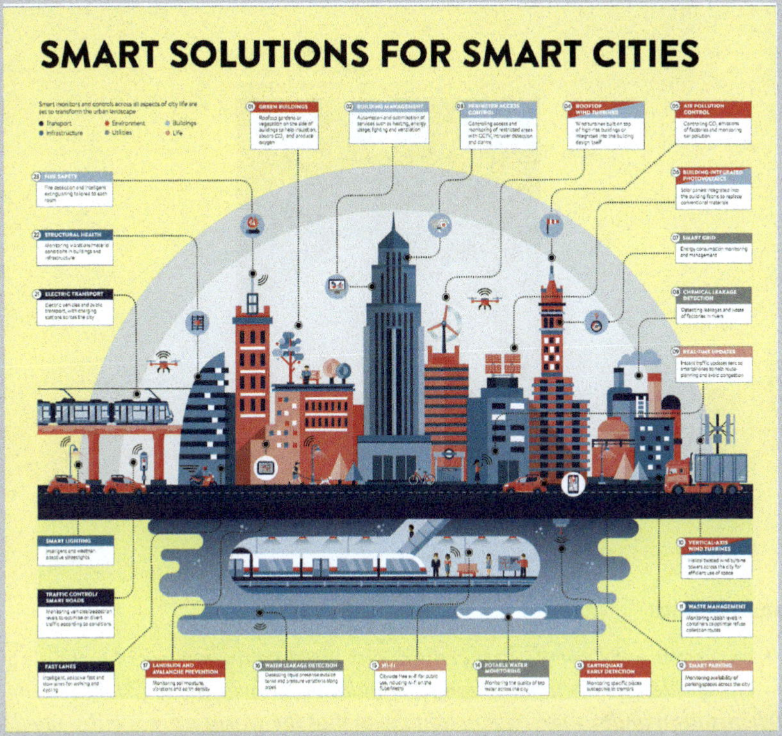

Smart Solutions for Smart Cities

A comprehensive example of smart city solutions including a range of smart monitors and controls across all aspects of city life, which are set to transform the urban landscape, based on six areas of transport, environment building, infrastructure, utilities, and life.

Available from: https://www.visualcapitalist.com/wp-content/uploads/2017/08/smart-cities.html.

Box 1.7 Examples of 'Smart Cities' reports and documents

Smart Cities

"Transforming the twenty-first century city via *the creative use of technology.*

The challenges of climate change, population growth, demographic change, urbanisation and resource depletion mean that the world's great cities need to adapt to survive and thrive over the coming decades.

There is an increasing interest, therefore, in the role that information and communications technologies could play in creating the cities of the future. But, as yet, few cities have fully grasped the possibility of becoming a 'smart city'.

Smart Cities outlines many of the opportunities for cities afforded by these contemporary technologies, indicating how the 'smart city' approach might fundamentally transform the way that cities are governed, operated, interacted with and experienced".

Available from: https://www.arup.com/perspectives/publications/research/section/smart-cities.

Box 1.8 Examples of 'Smart Cities' reports and documents

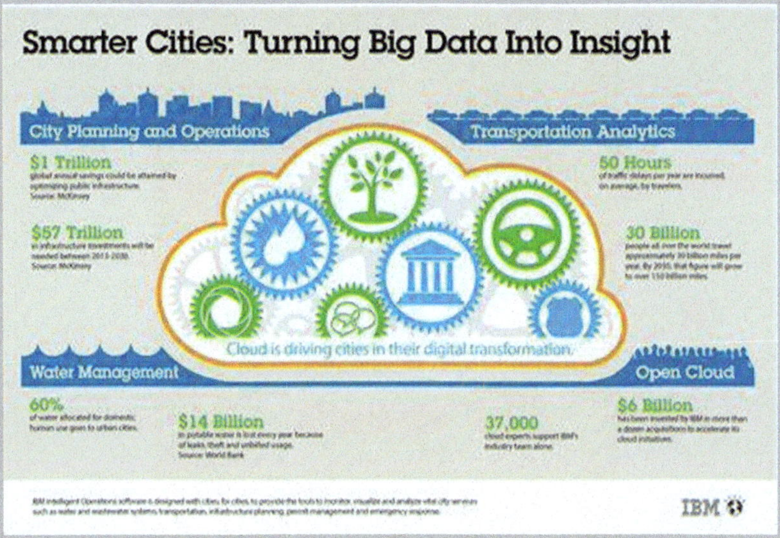

Smarter Cities: Turning Big Data into Insight

The comprehensive model developed by IBM showing four key areas for big data used in smart(er) cities, including:

(1) City Planning and Operations
(2) Transportation Analytics
(3) Water Management
(4) Open Cloud.

Similarly, Euroflash suggests a range of smart integration systems for city-wide platforms and cross domain applications, including smart energy, smart mobility, smart water, smart public services, smart buildings, and smart data center.

Available from: http://www-03.ibm.com/press/us/en/photo/42063.wss.

Box 1.9 Examples of 'Smart Cities' reports and documents

2015 Report

Big Data in Smart Buildings: Market Prospects 2015–2020

For years, the correlation between big data and smart buildings has been growing rapidly. There is a large market for this particular area at the building level.

"The report focuses on market sizing and opportunities for smart commercial buildings, providing a fresh market assessment based on original analysis and forecasts as well as a comprehensive analysis of the competitive landscape and crucial insights into what is driving M&A and investment in the market".

Available from: https://www.ifsecglobal.com/global/smart-buildings-and-big-data-2015-2020-research-download/.

Box 1.10 Examples of 'Smart Cities' reports and documents

2021 Report

Data for Better Lives

"World Development 2021: Data for Better Lives focuses on the potential of data to improve the lives of poor people, including through the creative use and re-use of data, and the essential elements of a data governance environment in the form of data infrastructure policy, the legal and regulatory framework, related economic policy implications, and institutional ecosystems. These diverse elements can be conceived of as the building blocks of a social contract that aims to deliver equitably on the potential benefits of data while safeguarding against harmful outcomes".

Available from: https://www.worldbank.org/en/events/2021/05/25/data-for-better-lives.

References

1. Aymen A, Kachouri A, Mahfoudhi A (2017) Data analysis and outlier detection in smart city. In: 2017 international conference on smart, monitored and controlled cities (SM2C), https://doi.org/10.1109/SM2C.2017.8071256. Also available from https://ieeexplore.ieee.org/document/8071256/

2. Bibri SE (2018) Smart sustainable cities of the future: the untapped potential of big data analytics and context–aware computing for advancing sustainability. Springer, Cham.https://doi.org/10.1007/978-3-319-73981-6_3
3. Cai Z, Kwak Y, Cvetkovic V, Deal B, Mörtberg U (2023) Urban spatial dynamic modeling based on urban amenity data to inform smart city planning. Anthropocene 42:100387. https://doi.org/10.1016/j.ancene.2023.100387
4. Cesario E (2023) Big data analytics and smart cities: applications, challenges, and opportunities. Frontiers Big Data 6. Section on Data Mining and Management. https://doi.org/10.3389/fdata.2023.1149402
5. Chen C-W (2023) Can smart cities bring happiness to promote sustainable development? Contexts and clues of subjective well-being and urban livability. Dev Built Environ 13:100108. https://doi.org/10.1016/j.dibe.2022.100108
6. Cheshmehzangi A (2022) ICT, cities, and reaching positive peace. Springer, Singapore
7. Cheshmehzangi A (2022b) The application of ICT and smart technologies in cities and communities: an overview. In: ICT, cities, and reaching positive peace. Springer, Singapore, pp 1–16
8. Cheshmehzangi A, Dawodu A, Sharifi A (2021) Sustainable urbanism in China. Routledge, New York
9. Cheshmehzangi A, Li Y, Li H, Zhang S, Huang X, Chen X, Su Z, Sedrez M, Dawodu A (2022) A hierarchical study for urban statistical indicators on the prevalence of COVID-19 in Chinese city clusters based on multiple linear regression (MLR) and polynomial best subset regression (PBSR) analysis. Sci Rep 12:1964. https://doi.org/10.1038/s41598-022-05859-8
10. Cheshmehzangi A, Su Z, Zou T (2023) ICT applications and the COVID-19 pandemic: impacts on the individual's digital data, digital privacy, and data protection. Frontiers Human Dyn 5. Section on Digital Impacts, Available fromhttps://doi.org/10.3389/fhumd.2023.971504
11. Davenport TH (2017) How analytics has changed in the last 10 years (and how it's stayed the same). Harvard Business Review: Analytics Services. Available from https://hbr.org/2017/06/how-analytics-has-changed-in-the-last-10-years-and-how-its-stayed-the-same
12. EPM Channel (2015) What is Big Data Discovery? Available from http://www.epmchannel.com/2015/04/07/what-is-big-data-discovery/
13. Jafari M, Kavousi-Fard A, Niknam T (2021) Stochastic synergies of urban transportation system and smart grid in smart cities considering V2G and V2S concepts. Energy 215(Part B):119054. https://doi.org/10.1016/j.energy.2020.119054
14. Kandt J, Batty M (2020) Smart cities, big data and urban policy: towards urban analytics for the long run. Cities 109:102992. https://doi.org/10.1016/j.cities.2020.102992
15. Kutty AA, Wakjira TG, Kucukvar M, Abdella GM, Onat NC (2022) Urban resilience and livability performance of European smart cities: a novel machine learning approach. J Cleaner Prod 378:134203. https://doi.org/10.1016/j.jclepro.2022.134203
16. Raven R, Sengers F, Spaeth P, Xie L, Cheshmehzangi A, de Jong M (2017) Urban experimentation and institutional arrangements. Eur Plann Stud 27(2). Urban Experimentation and Sustainability Transitions. https://doi.org/10.1080/09654313.2017.1393047
17. Ruchira G, Sengupta D (2023) Smart urban metabolism: a big-data and machine learning perspective. In: Bhadouria R, Tripathi S, Singh P, Joshi PK, Singh R (eds) Urban metabolism and climate change: perspectives for sustainable cities. Springer, Cham, pp 325–344. https://doi.org/10.1007/978-3-031-29422-8_16
18. Sagi A, Gal A, Czamanski D, Broitman D (2022) Uncovering the shape of neighborhoods: harnessing data analytics for a smart governance of urban areas. J Urban Manage 11(2):178–187
19. Shanmuganathan S (2014) From data mining and knowledge discovery to big data analytics and knowledge extraction for applications in science. J Comput Sci 10(12):2658–2665. https://doi.org/10.3844/jcssp.2014.2658.2665. Also available from https://thescipub.com/abstract/10.3844/jcssp.2014.2658.2665

20. Singh A, Kumar A (2022) Convergence of smart city with IoT and Big Data. Available from https://sciforum.net/paper/view/13930
21. Ushakov D, Dudukalov E, Mironenko E, Shatila K (2022) Big data analytics in smart cities' transportation infrastructure modernization. Transp Res Procedia 63(1):2385–2391. https://doi.org/10.1016/j.trpro.2022.06.274
22. Yao F, Wang Y (2020) Towards resilient and smart cities: a real-time urban analytical and geo-visual system for social media streaming data. Sustain Cities Soc 63:102448. https://doi.org/10.1016/j.scs.2020.102448

Chapter 2
Stock Market Prediction During COVID-19 Pandemic: A Time-Series Big Data Analysis Method

2.1 Introduction

The stock price is usually difficult to predict, especially during the COVID-19 pandemic. The pandemic has made stock prices more volatile, making it difficult for many investors to gauge trends, and has had a huge economic impact [1]. In this case, the need for stock price change forecasting has become more critical.

The US stock price dataset from Kaggle [2] and the COVID-19 data from World Health Organisation [3] give us an inspiration to find or propose a price pattern that matches the market during the pandemic in the US and then forecast future price changes. Promising solutions to model and forecast stock price pattern offer investors, manufactures, and even governments desirable benefits especially during COVID-19 pandemic [4].

There are several standard pricing patterns such as a cup with handle, and double bottom that are widely used to model and study the behaviour of stock prices [5, 6]. By realising this point, this chapter uses Dynamic Time Warping (DTW) algorithm to analyse and capture the similarities of the historical patterns and the COVID-19 stock pricing. This price pattern has the potential to help the future pandemic to monitor and manage stock markets. However, it is an analysis of the overall trend (i.e., the long-term trend) which is not always sufficient for investors to avoid the short-term risks. Hence, it is still required to put the stock price and COVID-19 data together and analyse them to understand how the outbreaks can affect the stock price.

There are several research projects that take the benefits of machine learning and computing solutions to predict the stock market. Jeon et al. [6] proposes an Artificial Neural Network (ANN) method based on selected features, whereas [7] builds a practical method that uses Structural Support Vector Machines (SSVMs) to handle the complex inputs to predict the future stock price trend. Ghosh and Chaudhuri [8] shows how stacking and deep neural network models can be deployed separately on feature-engineered and bootstrapped samples for forecasting stock trends during pre-pandemic and post-pandemic periods, while [9] proposes a federated learning framework to analyse and forecast stock market trends. These researches show remarkable

S. Pourroostaei Ardakani and A. Cheshmehzangi, *Big Data Analytics for Smart Urban Systems*, Urban Sustainability, https://doi.org/10.1007/978-981-99-5543-5_2

results; however, they focus more on the stock analysis and prediction in a usual period, and loss the attention of the influential variables [10] caused by COVID-19. In this study, a stock price forecasting method is proposed to extract and select data features from standard and COVID-19 stock pricing patterns, and construct a time-series machine learning model using long-short term memory (LSTM) [11] technique to predict the future stock price change. The proposed LSTM model is trained and tested using a public stock (i.e., Apple stock prices) and US's COVID-19 data.

This chapter is organised as: Sect. 2.2 reviews literature on the Big data analytics in stock prediction. Section 2.3 presents the research methodology keys, while Sect. 2.4 discusses the results. Section 2.5 summarises the discussions and outlines the findings.

2.2 Literature Review

Stock pattern recognition is one of the most promising technique to model and forecast stock pricing. By this, researchers and investors are fascinated in utilising this technique to capture similar pricing patterns and predict future stock trending [12]. For example, Fu et al. [13] proposes an approach to study and model time-series data patterns using Perceptually Important Points (PIP) technique. It also uses a time sliding window approach to reduce the dimensions of the time-series data. The results show that their PIP approach detects patterns earlier when compared with template matching methods. Kim et al. [14] utilises the DTW algorithm to recognise the market pattern. They applied this approach to a Pattern Matching Trading System (PMTS) for trading the index futures of the Korea Composite Stock Price Index (KOSPI 200). The system used the time series data from 9:00 am to 12:00 pm as the sliding window input and apply DTW to match and recognise those familiar patterns. Then, they use the patterns to predict the future price in the afternoon market on the same day. They claimed that their approach produces significant annualised returns. They also concluded that the patterns near the clearing time usually give more profits than other trading times.

Since the outbreak of COVID-19 in 2019, the world has changed, particularly the people's lives, world economy, and the fate of enterprises. Indeed, the pandemic has a significant impact on stock markets world-wide and the world coped with sharp stock price fluctuation due to the market uncertainty. Mathy [15] describes how the sudden stock price peaks can be caused by a variety of important factors - mainly stock availability, stability and demand. Goodell [16] suggests that COVID-19 could have a direct global devastating economic impact and unprecedented economic losses in the economy, banking and insurance, governments and public and financial markets. Yousfi et al. [17] confirm that COVID-19 has had a negative influence on the financial market and macroeconomics in the United States, and the pandemic's impact is more evident when economic uncertainty rises. Besides, Liu et al. [18] show that the announcement of the pandemic has had a significant negative impact on the worldwide stock market.

2.2.1 *Big Data Analytics in Stock Markets*

Big Data analytics is widely used to handle enormous amounts of stock data. However, it is still required to figure out and/or propose efficient feature extraction and machine learning techniques to improve the performance of data analysis and interpretation. For this, Shastri et al. [19] used Neural network models to predict stock market pricing on HIVE, while [20] prolepses a regression method to forecast stock price fluctuations on Spark. Ardakani et al. [9] builds a federated learning framework and uses Random forest, Support Vector Machine and Linear regression to predict stock pricing.

Stepwise regression is a regression method for feature extraction based on variable interpretation. This ensures the validity and significance of the selected variables while also reducing the additional error provided by redundant variables [21]. Indeed, it mainly solves the problem of multi-variable collinearity; that is, where the relationship between variables is not linear independent. It goes through each regression model step-by-step, with adding or removing feature variables based on stepping criteria [22]. There are three main approaches for stepwise regression, including forward selection, backward elimination, and bidirectional elimination [23]. For forward selection, it introduces features into the model one by one, adds feature variables that can improve the effect of the model on the selected set, and repeats this step until all variables are considered. For backward elimination, it starts by putting all features into the model, keeps the features that could lead to a significant change to the model, and repeats the process until no change. Bidirectional elimination is a combination method of the above two methods, it may perform forward selection and backward elimination for each adding feature. Stepwise analytic procedures are likely to be among the most widely used research techniques in both substantive and validity studies [24].

LSTM [11] is a variant of SimpleRNN [25], it is a recurrent network with an innovative design to handle long-term temporal sequence [26]. It informatively adds a mechanism to remember the last trained result, i.e., a carry. A carry stores the historical output of each RNN node, and updates itself for each iteration of training. There is also a "forget gate" [27] for dropping out legacy memory so that the network does not meet gradient explosion after a large number of iterations.

The input is a sequence of temporal-connected features and one training label. Figure 2.1 for example, each block represents an item of input, which includes both features $f_1, ..., f_n$ and their coordinating labels $label_1, ..., label_n$. In this data frame of 7 items, the first 6 items are used for training, while the 7_{th} is set to be the label of this data frame.

In 2017, Roondiwala [29] and the authors proved LSTM is able to predict stock price with features such as open, high, low, and close (OHLC). They applied their model to predict the stock return of Nifty. They also gave an explanation of the mechanism of RNNs to predict this kind of time-series data. They concluded that the

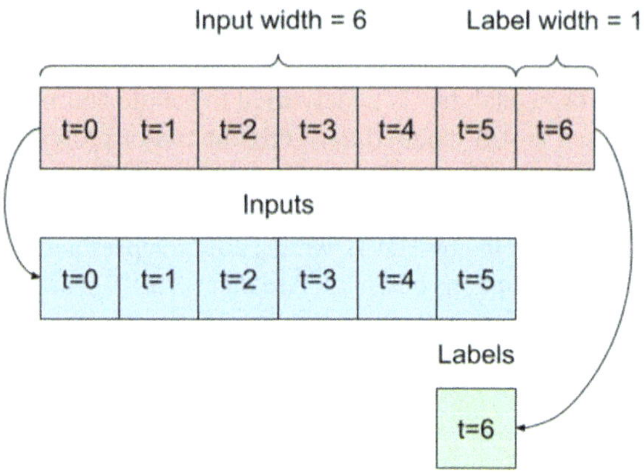

Fig. 2.1 LSTM input data [28]

LSTM has a good performance with the prediction errors RMSE only 0.00859 for the testing data due to the hidden dynamics of the RNN. In the same year, [30] used different types of recurrent neural networks (RNN) to predict Google's stock price. They tried three types of RNN models, including Gated Recurrent Units (GRUs), basic RNN, and LSTM models. They compared the performance of the machine learning models to figure-out the best-fitted one for Google's stock price prediction. It concluded that the proposed LSTM model outperforms the other ones. Shah et al. [31] used LSTM to predict market movements. The results supported that LSTM technique gives a higher accuracy in both daily and 7-day predictions. They also used larger datasets (20 years data) to train the model and and showed that LSTM works better with larger datasets since there are more data records, features and fluctuations to model/recognise the patterns.

2.3 Methodology

This study uses several important steps to predict stock prices during the COVID-19 pandemic. As Fig. 2.2 shows, the proposed data analytics approach starts with a data preprocessing step. Then, a feature extraction technique (stepwise regression) is used to select meaningful features, and a DTW technique is used to capture stock price pattern similarities. This forms a predictive feature pool to train the LSTM model. Tensorflow libraries are used to build the feature extraction approach, and train and test the machine learning model. Yet, both the datasets (Apple stock and COVID-19) are built using Hadoop Distributed File System which interacts with pyspark scripts in YARN mode.

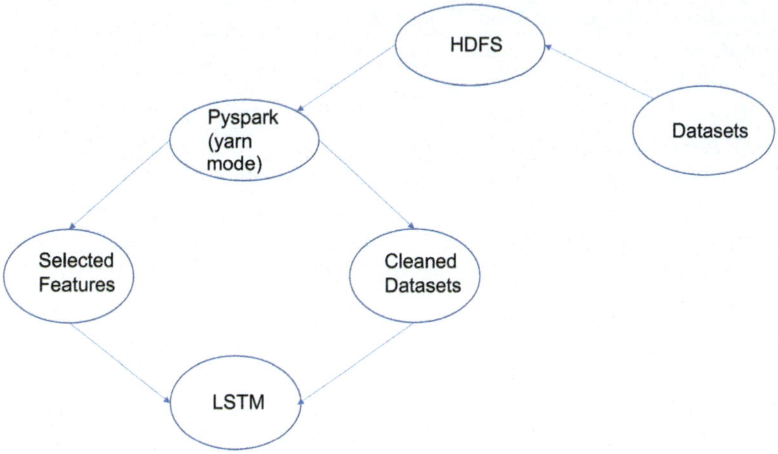

Fig. 2.2 Architecture of the proposed method

2.3.1 Data Preprocessing

This study uses two public datasets: (1) stock daily data (i.e., Apple stock), and (2) US's COVID-19. The former includes the historical daily stock data such as trading date, close price, traded volume, and so forth, while the latter gives information of US states' COVID-19 cases and death numbers. A data cleaning approach is also used to deal with null, missing or negative values, format alignment, filter uncorrelated features and normalised data. SparkSQL starts to manage the data cleaning and preprocessing once the datasets are converted into Spark Dataframes.

2.3.2 Pattern Retrieval Using DTW

This is the key to figure out and study the similarities of stock market patterns (mainly stock prices) to compare and model their behaviours. For this, we used the stock patterns of the first 365 d during the COVID-19 period, and built the patterns from a predefined day. A sliding window method is used to compare the historical and the current patterns. DTW technique (see Algorithm1) is used to compare the similarity of the two time-series patterns. DTW measures the similarities of time series data with different lengths, and is widely used in pattern matching [32].

Algorithm 1: Dynamic Time Warping

Input: two arrays (time series) ts_a and ts_b

Output: DTW distance between two time series

```
d ← λ x, y: ||x − y||
M, N ← ts_a.length, ts_b.length
cost = Matrix_{M*N} where each element is inf value
cost[0][0] ← d(ts_a[0], ts_b[0])
for i = 1 → M do
    cost[i][0] ← cost[i-1][0] + d(ts_a[i], ts_b[0])
end for
for j = 1 → N do
    cost[0][j] ← cost[0][j-1] + d(ts_a[0], ts_b[j])
end for
max_warping_window gets 10000
for i = 1 → M do
    for j = max(1, i- max_warping_window) →
    min(N, i + max_warping_window)
        choices ← (cost[i-1][j-1], cost[i][j-1], cost[i-1][j])
        cost[i][j] ← min(choices) + d(ts_a[i], ts_b[j])
    end for
end for
return cost[-1][-1]
```

2.3.3 Feature Selection

Feature selection plays a key role in machine learning model training particularly supervised learning. There are many feature extraction methods aiming to extract meaningful data features, and reduce dataset dimension. The "close price" is the target feature, and the input data is transformed into DataFrame having features and labels. A greedy stepwise regression algorithm is used to select the optimal feature subset. Equation 2.1 shows the normalisation method, min-max, which is used to normalise the data features [0, 1], where X represents the feature column.

$$F(X_i) = \frac{X_i - \min(X_i)}{\max(X_i) - \min(X_i)}. \tag{2.1}$$

$$R^2 = 1 - \frac{\sum(y - \widehat{y})^2}{\sum(y - \overline{y})^2} \tag{2.2}$$

This study utilises Algorithm 2 to run a linear regression technique on each feature separately to see the highest correlation with the target feature. The feature with the highest R-squared value (see Eq. 2.2) is the first selected feature. This is repeated to calculate the R-squared values for other features until the best result and the most-fitted predictive vector are achieved. According to the results, R-squared result is increased if the feature (or the feature vector) is meaningful for the target prediction, while this is reduced if the feature is not correlated to the target.

Algorithm 2: Feature Selection

```
Input: normalized features, close price label
Output: selected features
Model ← LinearRegression(features without highly related one)
for i = 1 → len(features) - 1 do
    Loss ← argmin Model(features[i])
end for
Get the best feature with minimum Loss
Loss-Select ← argmin Model(best feature)
Selection Set ← best feature
features.Remove(best feature)
for i = 1 → len(features) - 1 do
    Selection Set.Add(features[i])
    features.Remove(features[i])
    Model ← LinearRegression(Selection Set)
    Loss-Select-New ← argmin Model(Selection Set)
    if Loss-Select-New ≤ Loss-Select
        Loss-Select ← Loss-Select-New
        continue
    else
        Selection Set.Remove(features[i])
    end if
end for
return Selection Set
```

Two most similar historical patterns of the regression analysis are used to form the optimal feature subset. As Algorithm 2 shows, each similar pattern gives us an optimal feature vector. Hence, the intersection of both the feature vectors (two similar patterns) should be given to train the prediction model.

2.3.4 Predicted Stock Data Generation Using LSTM

Using Stepwise Regression, a set of correlated features is selected to train a LSTM network. LSTM is a variant of SimpleRNN (Recurrent Neural Network), which links the output of each RNN layer to the input of the next network layer. It saves past information through time-steps and allows to extract time-series features over long periods. Hence, the features of the last 10 d are stored as a Numpy array, while the close-price of 11_th day is set as the target.

A min-max scaler technique is used to normalise the time-series data. As Eqs. 2.3 and 2.4 show, this technique maps the values to the range of 0 and 1. The normalized data is fed into the neural network, and the output data is converted to the original scale when extracting the final result.

$$X_{std} = \frac{X - X.\min(axis = 0)}{X.\max(axis = 0) - X.\min(axis = 0)} \tag{2.3}$$

$$X_{norm} = X_{std} * (\max - \min) + \min \tag{2.4}$$

The LSTM network is built using the following script on Tensorflow [33].

```
grid_model = Sequential()
grid_model.add(LSTM(50, return_sequences=True,
        input_shape=(10,5)))
grid_model.add(LSTM(50))
grid_model.add(Dropout(0.2))
grid_model.add(Dense(1))
```

2.4 Result Analysis and Discussion

2.4.1 Data Preprocessing

Data preprocessing aims to convert the stock and COVID-19 datasets into the pyspark dataframe and check the dataset's null or missing values. Tables 2.1 and 2.2 show a sample of AAPL and US's COVID-19 data.

Table 2.1 AAPL statistics

	Date	Volume	High	Adjusted close
Count	10416	10416	10416	10416
Mean	/	332010624	14.22	13.46
Standard deviation	/	339282074	30.80	30.20
Min	1/2/1982	0.00	0.05	0.04
Max	31/12/2021	7421640800	182.94	181.78

Table 2.2 Covid-19 statistics in US

	Date reported	Country code	Cumulative cases	Cumulative deaths
Count	841	841	841	841
Mean	/	/	27500205	445014
Stardard deviation	/	/	24426545	309812
Min	3/1/2020	US	0	0
Max	22/4/2022	US	80006661	982322

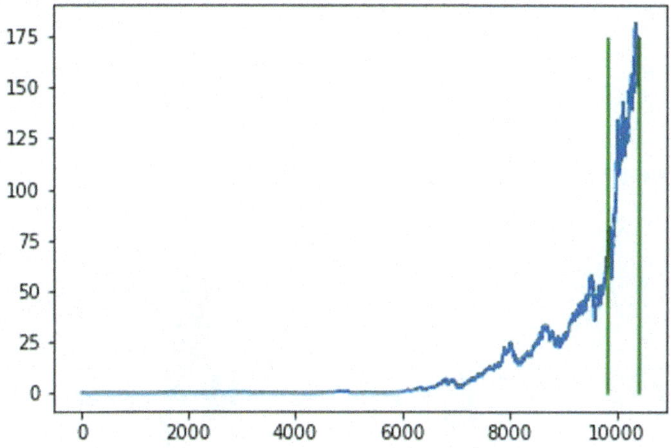

Fig. 2.3 AAPL's stock price

Figure 2.3 shows the AAPL's stock price (from day 0 to day 10410) where the COVID-19 period is depicted by the green straight lines. For this, the trading date is mapped between the first (day 0) and the last trading dates (day 10410) in the stock data, while the COVID-19 period is between day 9848 and day 10410.

2.4.2 Estimation of Close Price and COVID-19 Data

The behaviour of close price is analysed against the deaths and new cases during the pandemic to highlight the relationship between the stock price and COVID-19. Figures 2.4 and 2.5 shows the stock trending where the pandemic leads to an upward trend of the AAPL stock. For example, the price consistently is increased when the number of new cases or new deaths enhanced.

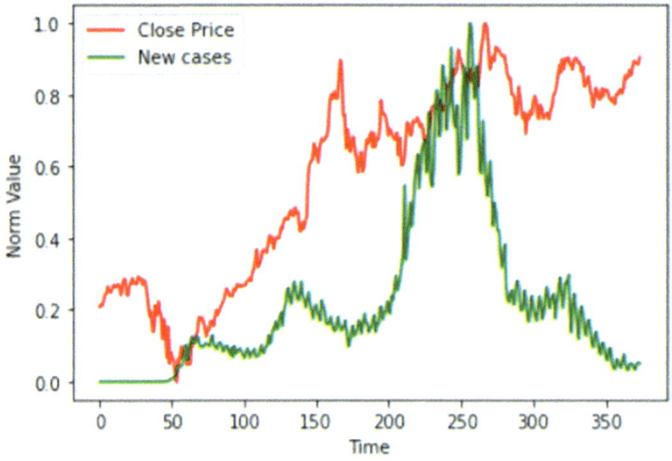

Fig. 2.4 Close price and new cases

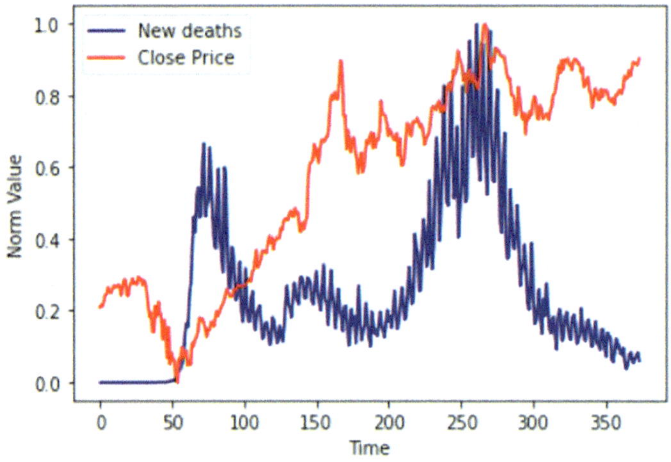

Fig. 2.5 Close price and new deaths

2.4.3 Pattern Selection

A DTW-based distance is calculated between the current and historical patterns after generating the historical patterns using the sliding window method. Then, the top two patterns are selected to have the smallest DTW distance. Figure 2.6 shows the results where the data patterns are similar between the two red lines.

Figure 2.7 shows the general trends of the COVID-19 data and the most similar stock pattern. According to this, both patterns are recognised as a standard data pattern, called cup with a handle pattern. Hence, it is concluded that AAPL pricing behaves like the cup with handle data pattern if any pandemic occurs in the future.

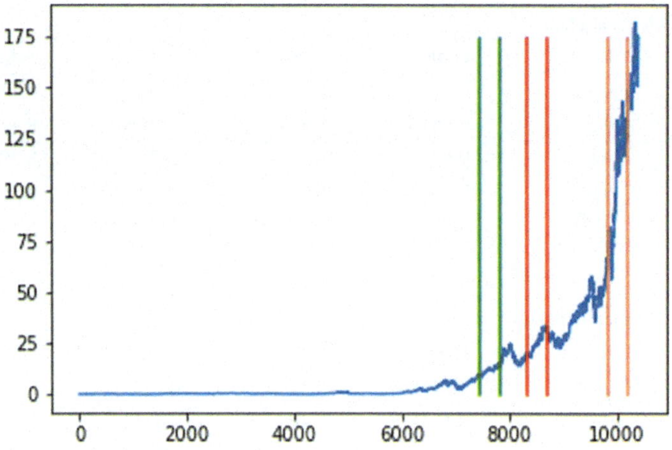

Fig. 2.6 Top 2 similar patterns

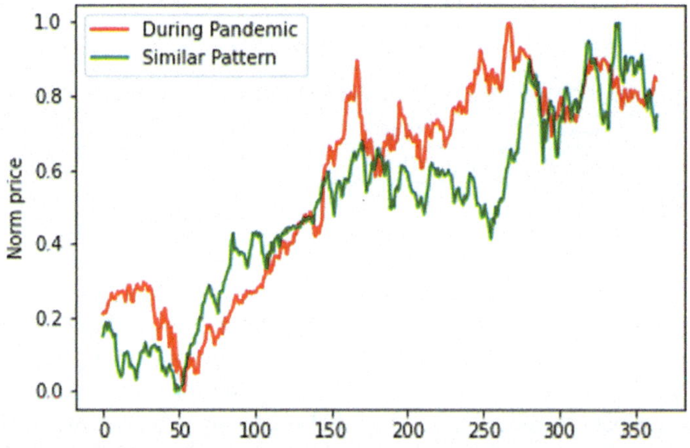

Fig. 2.7 The COVID-19 data and the most similar pattern

2.4.4 Feature Selection Result with Analysis

To make the linear regression fitting in feature selection smoother, all stock data features in AAPL should be normalized in advance. Tables 2.3 and 2.4 shows a five day-sample of the originak and normailzied features of AAPL stock.

A linear regression model is trained to select the best fitted features. Moreover, R-squared is measured to see the feature set which minimises the prediction errors using the linear regression predictions. Table 2.5 shows the best result and the most similar perdition pattern where"High" (means high price) feature with the highest R-squared value is selected as the first predictive feature.

Table 2.3 Original Data

	Low	High	Open	Volume	Adjusted close
1	9.31	9.56	9.32	783678000	8.17
2	9.63	9.75	9.66	872855200	8.31
3	9.69	9.82	9.72	784621600	8.38
4	9.60	9.96	9.92	776490400	8.26
5	9.70	9.86	9.72	717262000	8.37

Table 2.4 Normalized Data

	Low	High	Open	Volume	Adjusted close
1	0.1389	0.1366	0.1255	0.3633	0.1499
2	0.1881	0.1645	0.1772	0.4151	0.1752
3	0.1987	0.1759	0.1862	0.3639	0.1873
4	0.1838	0.1977	0.2158	0.3591	0.1659
5	0.1992	0.1812	0.1857	0.3248	0.1861

Table 2.5 R-squared for Selecting the Most Similar Pattern

	Low	High	Open	Volume
R-squared	0.8992	0.9433	0.8364	−25.8153

Figure 2.8 shows the feature selection process results. According to this, a feature should be added to the predictive feature set if this improves the R-squared. After examining all the original features, a set of five features including "Low", "Open", "Volume", "High", and "Adjusted Close" is selected. A similar approach is used for the second pattern and by this, a list of "Low", "Open", "Volume", "High", and "Adjusted Close" features are selected. Finally, both the feature sets are combined and the intersected features including'Low", "Open", "Volume", "High", and "Adjusted Close" are used as the predictive feature sets to build the LSTM model.

2.4.5 Result for LSTM Price Prediction

KerasRegressor is used on a Tensorflow platform to build and test the model and find the best-fitted training parameters. For this, Adam and Adadelta are selected as the optimizer, with batch size of 16 and 20, and 50 and 100 training epochs. As Table 2.6 shows, the proposed model returns the best training results when the optimizer is Adam, the batch size is set to 20, and training of 100 epochs.

As for the validation, the test dataset is scaled to a normalize interval between 0 and 1. The model is then used to predict the input features. However, converting the

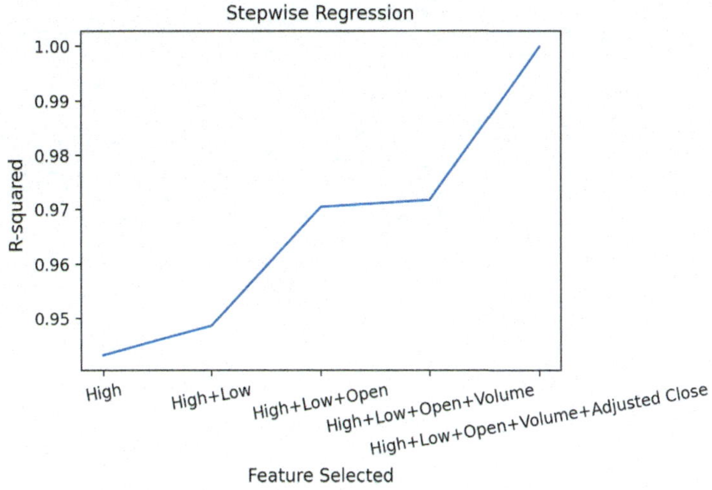

Fig. 2.8 Feature selection process

Table 2.6 Training evaluation on best parameter

Optimizer	Batch size	Epoch	Loss	Validation loss
Adam	20	100	0.0062	0.0330

outputs to the original price range is necessary while the result retrieval. As Fig. 2.9 shows, our model is well fitted for stock price prediction. This model correctly predicts the stock trending and general patterns. It returns an RMSE of 85.19 during the validation test that gives a good prediction accuracy.

2.4.6 Predicted Price and Covid-19 Data Factors

The relationship between the predictions (stock price) and the COVID-19 data (new deaths and cases) are studied to evaluate the accuracy of the trained model. As Figs. 2.10 and 2.11 show, there are similar behaviours between the stock and COVID-19 data patterns. Figure 2.10 depicts the overall data trend of the predicted stock price and the COVID-19 data that have similar patterns, while Fig. 2.11 shows two pandemic waves, as the blue boxes, that give the number of new deaths and reported cases. According to this, there is a gradual downward trend after rising to the local peak for both the stages. This matches the stock price prediction results where a peak is reached when the number of new deaths and reported cases are increased and reach a peak, but then they are gradually decreased.

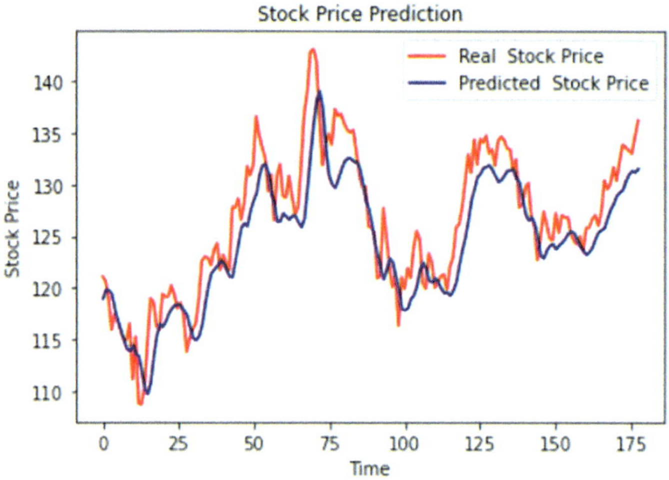

Fig. 2.9 Price prediction

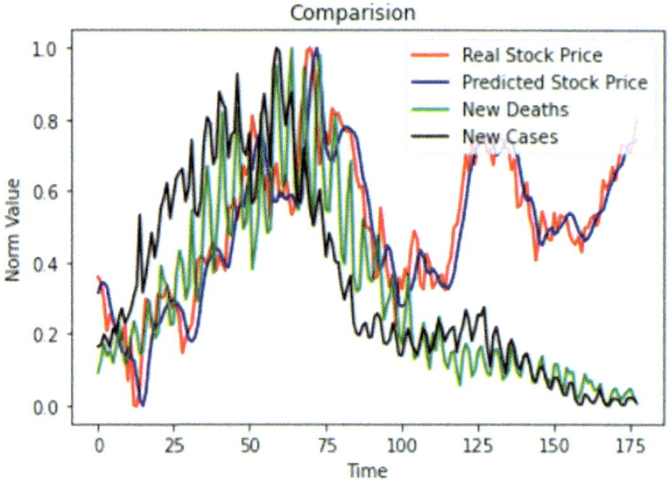

Fig. 2.10 Covid-19 data and stock price

2.5 Conclusion

This chapter studies the behaviours of stock markets during the COVID-19 outbreak and proposes a Big data-enabled machine learning approach to predict the prices and market trending for any future pandemic. For this, the behaviour of Apple stock data is analysed to model stock pricing during the COVID-19 pandemic. Hence, the stock data patterns are identified to find deeper connections between stock data and pandemic data. A data pre-processing approach is used to clean and prepare two

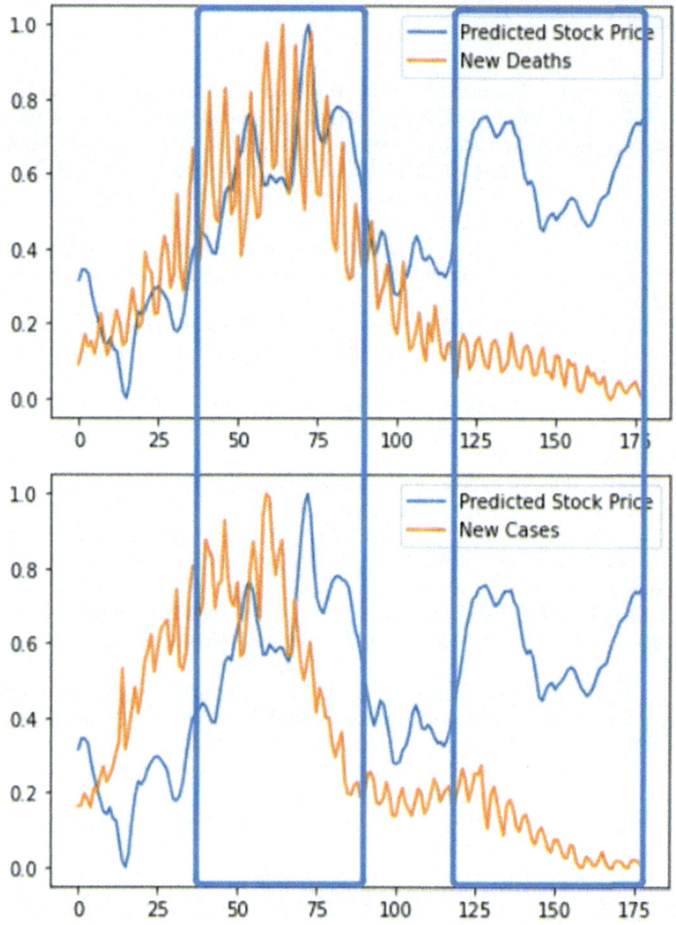

Fig. 2.11 Similar trend

big and public datasets, while a feature extraction technique (stepwise regression) is used to select the meaningful. Moreover, a DTW technique is used to capture stock price pattern similarities and LSTM machine learning model is trained and tested to predict the stock prices in the case of any future pandemic. According to the results, a standard data pattern, a cup with handle, is recognised for the behaviour of stock data during the pandemic. Hence, it is concluded that there is a certain connection between stock data during the epidemic and stock data during regular days.

Acknowledgements The research work in this chapter was supported by our research team members: Chen XU, Chenglei You, Donglin Jiang, and Pui Yee CHENG

References

1. Kaye AD, Okeagu CN, Pham AD, Silva RA, Hurley JJ, Arron BL, Sarfraz N, Lee HN, Ghali GE, Gamble JW, Liu H, Urman RD, Cornett EM (2021) Economic impact of covid-19 pandemic on healthcare facilities and systems: international perspectives. Best Pract Res Clin Anaesthesiol 35(3):293–306. ISSN 1521-6896. https://doi.org/10.1016/j.bpa.2020.11.009
2. Mooney P (2023) Stock market data-kaggle. https://www.kaggle.com/datasets/paultimothymooney/stock-market-data. Retrieved March 2023
3. WHO (2023) Global research on coronavirus disease (covid-19). https://www.who.int/emergencies/diseases/novel-coronavirus-2019/global-research-on-novel-coronavirus-2019-ncov. Retrieved March 2023
4. Mottaghi N, Farhangdoost S (2021) Stock price forecasting in presence of covid-19 pandemic and evaluating performances of machine learning models for time-series forecasting. arXiv preprint arXiv:2105.02785
5. Jeon S, Hong B, Lee H, Kim J (2016) Stock price prediction based on stock big data and pattern graph analysis. In: International conference on internet of things and big data, vol 2. SCITEPRESS, pp 223–231
6. Jeon Seungwoo, Hong Bonghee, Chang Victor (2018) Pattern graph tracking-based stock price prediction using big data. Future Gen Comput Syst 80:171–187
7. Leung CK-S, MacKinnon RK, Wang Y (2014) A machine learning approach for stock price prediction. In: Proceedings of the 18th international database engineering and applications symposium, pp 274–277
8. Ghosh I, Chaudhuri TD (2021) Feb-stacking and feb-DNN models for stock trend prediction: a performance analysis for pre and post covid-19 periods. Decis Mak Appl Manag Eng 4(1):51–84
9. Ardakani SP, Du N, Lin C, Yang J-C, Bi Z, Chen L (2023) A federated learning-enabled predictive analysis to forecast stock market trends. J Ambient Intell Human Comput
10. Nivoix S, Rey S (2021) Covid-19: stock markets responses. Ideas 06
11. Staudemeyer R, Morris E (2019) Understanding LSTM – a tutorial into long short-term memory recurrent neural networks. Neural Evolutionary Comput 09
12. Brunelli R, Poggiot T (1997) Template matching: matched spatial filters and beyond. Pattern Recogn 30(5):751–768. ISSN 0031-3203
13. Fu T, Chung F, Luk R, Ng C (2005) Preventing meaningless stock time series pattern discovery by changing perceptually important point detection. In: International conference on fuzzy systems and knowledge discovery
14. Kim SH, Lee HS, Ko HJ, Jeong SH, Byun HW, Oh KJ (2018) Pattern matching trading system based on the dynamic time warping algorithm. Sustainability 10(12)
15. Mathy Gabriel P (2016) Stock volatility, return jumps and uncertainty shocks during the great depression. Financ Hist Rev 23(2):165–192. https://doi.org/10.1017/S0968565016000111
16. Goodell JW (2020) Covid-19 and finance: agendas for future research. Financ Res Lett 35:101512
17. Yousfi M, Zaied YB, Cheikh NB, Lahouel BB, Bouzgarrou H (2021) Effects of the covid-19 pandemic on the us stock market and uncertainty: a comparative assessment between the first and second waves. Technol Forecast Social Change 167:120710
18. Liu M, Choo W-C, Lee C-C (2020) The response of the stock market to the announcement of global pandemic. Emerg Market Financ Trade 56(15):3562–3577
19. Shastri M, Roy S, Mittal M (2019) Stock price prediction using artificial neural model: an application of big data. EAI Endors Trans Scalable Inf Syst 6(20)
20. Awan MJ, Rahim MFM, Nobanee H, Munawar A, Yasin A, Zain AM (2021) Social media and stock market prediction: a big data approach; Awan MJ, Shafry M, Nobanee H, Munawar A, Yasin A, et al (2021) Social media and stock market prediction: a big data approach. Comput Mater Continua 67(2):2569–2583
21. Tian Yu, Guang Yu, Li Peng-Yu, Wang Liang (2014) Citation impact prediction for scientific papers using stepwise regression analysis. Scientometrics 101(2):1233–1252

22. Ruengvirayudh P, Brooks GP (2016) Comparing stepwise regression models to the best-subsets models, or, the art of stepwise. General Linear Model J 42(1):1–14
23. Wahono B, Ogai H (2012) Construction of response surface model for diesel engine using stepwise method. In: The 6th international conference on soft computing and intelligent systems, and the 13th international symposium on advanced intelligence systems, pp 989–994. https://doi.org/10.1109/SCIS-ISIS.2012.6505171
24. Thompson B (1989) Why won't stepwise methods die?
25. Tjandra A, Sakti S, Nakamura S (2017) Compressing recurrent neural network with tensor train. Mach Learn 05
26. Cho K, Merrienboer B, Gulcehre C, Bougares F, Schwenk H, Bengio Y (2014) Learning phrase representations using RNN encoder-decoder for statistical machine translation. Comput Lang 06. https://doi.org/10.3115/v1/D14-1179
27. Gers F, Schmidhuber J, Cummins F (2000) Learning to forget: continual prediction with LSTM. Neural Comput 12:2451–71. https://doi.org/10.1162/089976600300015015
28. TenserFlow (2023) Tensorflow keras offical document. https://keras.io/api/layers/recurrent_layers/lstm/. Retrieved March 2023
29. Roondiwala M, Patel H, Varma S (2017) Predicting stock prices using LSTM. Int J Sci Res (IJSR) 6:04. https://doi.org/10.21275/ART20172755
30. Honchar Oleksandr, Di Persio Luca (2016) Artificial neural networks approach to the forecast of stock market price movements. Int J Econo Manag Syst 1(158–162):01
31. Shah D, Wesley C, Farhana Z (2018) A comparative study of LSTM and DNN for stock market forecasting. In: IEEE international conference on big data (big data)
32. Berndt DJ, Clifford J (1994) Using dynamic time warping to find patterns in time series. In: KDD workshop, vol 10. Seattle, WA, USA, pp 359–370
33. Tensorflow. https://www.tensorflow.org/. Retrieved March 2023

Chapter 3
A Big Data Solution to Predict Cryptocurrency Market Trends: A Time-Series Machine Learning Approach

3.1 Introduction

Cryptocurrencies are digital or virtual assets that are secured using computer cryptography algorithms based on decentralized computer platforms-mainly blockchain [9]. They (e.g., Bitcoins) have the capacity to offer economic benefits such as security and time/cost efficiency. A blockchain deploys data blocks on a decentralised computing platform to store and process data [21]. Each block contains data attributes (i.e., transactions) that are processed and independently verified by each network member. It allows cryptocurrency transactions to be permanently recorded in distributed ledgers focusing on the characteristics of the blockchain platform and the venture capital [34].

The cryptocurrency market has been rapidly growing over the past years due to the cryptocurrency's ease of use and reliability [33]. Economists and market investors analyse and forecast market behaviour/trend to enhance market profits. However, it is still required to propose an accurate and efficient time-sensitive prediction approach to model market behaviour. Machine learning techniques have the capacity to analyse and predict cryptocurrency market trends/behaviours.

This chapter proposes a predictive analysis approach using machine learning to model and forecast cryptocurrency market trends. It uses a public time-series dataset [2]. This dataset is prepared via a three-step data pre-processing approach which manages the missing values, normalizes data and converts time-series samples to meet supervised machine learning requirements. This research builds a Dynamic Time Wrapping (DTW) K-means clustering approach to recognise cryptocurrency data patterns. Moreover, two machine learning models, including Long-Short Term Memory (LSTM) [12] and Random Forest (RF) [8] are trained to forecast cryptocurrency (i.e., Bitcoin) price. They are tested and evaluated in terms of Mean Absolute Error, Mean Squared Error, Root Mean Squared Error, and R-squared scores to figure out the best-fitted solution. This research offers the following contributions:

S. Pourroostaei Ardakani and A. Cheshmehzangi, *Big Data Analytics for Smart Urban Systems*, Urban Sustainability, https://doi.org/10.1007/978-981-99-5543-5_3

- To propose a three-step data processing approach to prepare a multi-variate time-series cryptocurrency dataset for supervised learning data analysis.
- To analyse cryptocurrency price trends and recognise data patterns using a time-series clustering approach.
- To build, compare and evaluate two machine learning models for cryptocurrency trend prediction.

The remainder of this chapter is organized as follows. Section 3.2 reviews literature, introduces relevant state-of-the-art cryptocurrency trend analysis solutions, and highlights the existing research gaps. Section 3.3 presents the research methodology and the proposed approach. Section 3.4 presents and discusses the experimental results. Section 3.5 summarises the key findings of this research and addresses future works.

3.2 Literature Review

This section summarises the literature on cryptocurrency market prediction. For this, data pattern clustering methods and cryptocurrency (i.e., Bitcoin) price prediction approaches are discussed and explained to highlight their similarities and differences.

3.2.1 Cryptocurrency Pattern Recognition and Clustering

There are several statistical and economic studies focusing on cryptocurrency data pattern modelling and the market price behaviour recognition [6]. Cryptocurrency price clustering patterns and synergies are varying and rapidly changing [20]. As [32] reports, the behaviour of the cryptocurrency market is highly correlated to speculation and investment. It can be influenced by social media and networks-mainly Twitter especially during the bubble price period [7, 26].

Cryptocurrency market analysis needs time-series data processing approaches to study and recognise the data patterns [31]. Chan et al. [5] aimed to compare and evaluate the performance of distance measurement approach via a time-series clustering method (K-medoids) to predict Bitcoin price. However, there is a complex micro-structural behaviour in the market due to the tail behaviour of the log-return distribution of cryptocurrencies [25]. It needs to use a convergence approach to integrate the various behaviours of the cryptocurrencies to be persistent and lasting [3, 18]. Ozer and Sakar [23] designed an Automated Cryptocurrency Trading System model to avoid the possible risks of cryptocurrency transactions, while [30] proposed a real-time data analysis approach to integrate Bitcoin price behaviours and detect the price anomalies and bubble crashes.

3.2.2 *Bitcoin Price Prediction*

Machine learning techniques have the capacity to analyse and predict cryptocurrency behaviours and trends. Shah and Kang [29] devise a Bitcoin trading strategy and report the efficacy of the Bayesian regression method for predicting the price variation of Bitcoin. However, this approach is not verified in large-scale applications with high volume investments as it needs to build a scalable computation architecture. Kolla [16] uses a Linear Regression model to train several important features for Bitcoin price prediction. It also predicts the price of Bitcoin using a Recurrent Neural Network (RNN) with Long Short Term Memory (LSTM) cells. Hotz-Behofsits et al. [13] propose a medium-scale multivariate state-space model to match the pronounced degree of volatility. They figure out that introducing suitable shrinkage priors can allow for time-varying parameters and error distribution. Guo et al. [10] propose a temporal mixture model to figure out the dynamic effect of order features on the volatility evolution. Compared with the basic regression and ensemble models, their mixture models and XGT methods are more robust in the time-varying environment of the Bitcoin market. Naimy and Hayek [22] report that the GARCH method has a better predictive ability when the volatility of Bitcoin is relatively low. They show that the behaviour of Bitcoin is not as same as the behaviour of other cryptocurrencies. Jalali and Heidari [14] use grey system theory to predict Bitcoin with a small number of data and incomplete information. The result of these predictions shows that the price of Bitcoin can be predicted using a 5-days time window with an average error rate of 1.14%. Septiarini et al. [28] construct the classic statistic and artificial intelligence models to predict Bitcoin price. They compare the performance of RMSE and MSE. LSTM is another machine learning technique that is suitable for processing and predicting important events with very long intervals and delays in time series. As [11] reports, it has a better performance in cryptocurrency data prediction as compared to RNN and HMM .

According to the literature review, there is still a gap in proposing a predictive machine learning solution to recognise Bitcoin data patterns and forecast the behaviours.

3.3 Methodology

This section introduces the research methodology focusing on Bitcoin price pattern recognition and behaviour prediction. For this, a three-step data pre-processing approach is used to clean and prepare a big Bitcoin price dataset for machine learning analysis and further processing. Moreover, a time-series clustering technique is proposed to cluster and recognise Bitcoin price patterns, while two machine learning approaches are trained and evaluated to predict price behaviours.

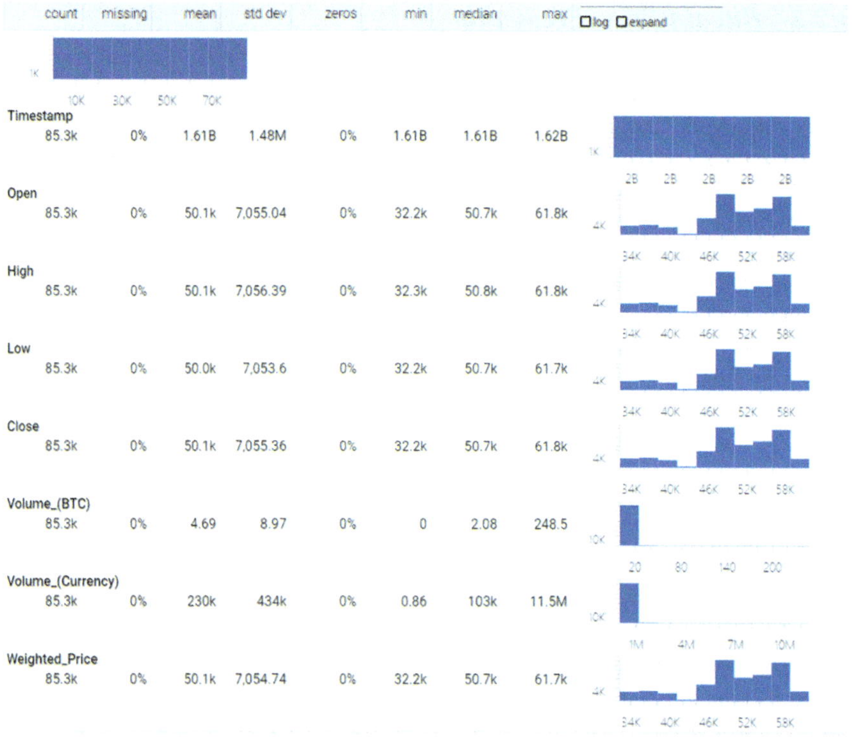

Fig. 3.1 Dataset overview after null value removal

3.3.1 *Dataset Selection and Pre-processing*

This study uses a public dataset containing 23 CSV files, containing a total of about 37,000 lines of cryptocurrency data recorded from April 28, 2013, until June 19, 2021 [2]. Each CSV file contains 10 features, namely SNo (Series Number), Name of the cryptocurrency, Symbol, Date, Open-price, Close-price, High-price, Low-price, Volume of Transactions, and Market capitalization in USD [35].

Data pre-processing contains three steps as below:

- Handling missing data: there are two main approaches, including (1) discarding any samples that contain a missing value, and (2) updating the null value with an aggregated score (e.g., column's mean or median) to deal with missing data. This paper removes the missing values that only take a minimal percentage of the whole dataset. Figure 3.1 shows the data distribution after null value removal.
- Normalization: it is used when data features are in various ranges. It may influence the machine learning predictions, and result in misclassification. Normalisation techniques transform the values to a normal scale (e.g., 0–1) without distorting

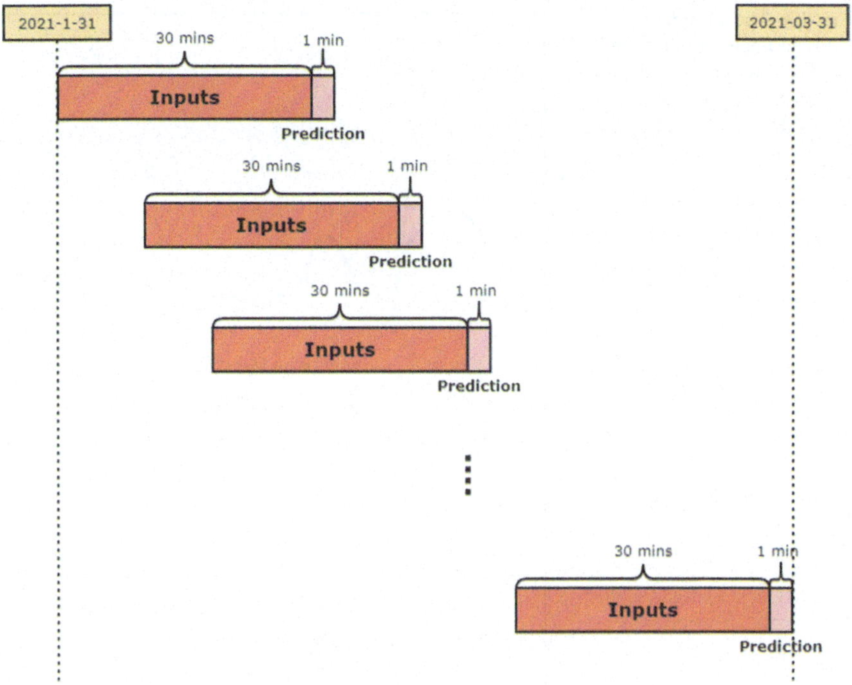

Fig. 3.2 Convert time-series dataset to supervised machine learning

differences in the value ranges. A MinMaxScaler is used in this research to scale the values between 0 and 1.

- Converting time-series data: supervised machine learning techniques can be used to handle time-series data predictions if the dataset is re-formed as framed data. Figure 3.2 shows a sliding window approach that is used to transform time-series data to meet supervised machine learning requirements.

3.3.2 Data Pattern Recognition via Clustering

Time-series clustering is an estimated calculation based on similarity or distance rather than perfect matching in traditional clustering methods [36]. There are three main distance measures: change, time, and shape similarities. The comparison results of the time-series data are low if the datasets have a large estimated distance.

Cryptocurrency data depends on several external factors (e.g., market owners, investors, and social media) and suffers from the lack of a specific pricing agreement. Hence, traditional distance measurement methods such as Euclidean distance may fail to recognise the data pattern similarities due to the lack of an equal data length

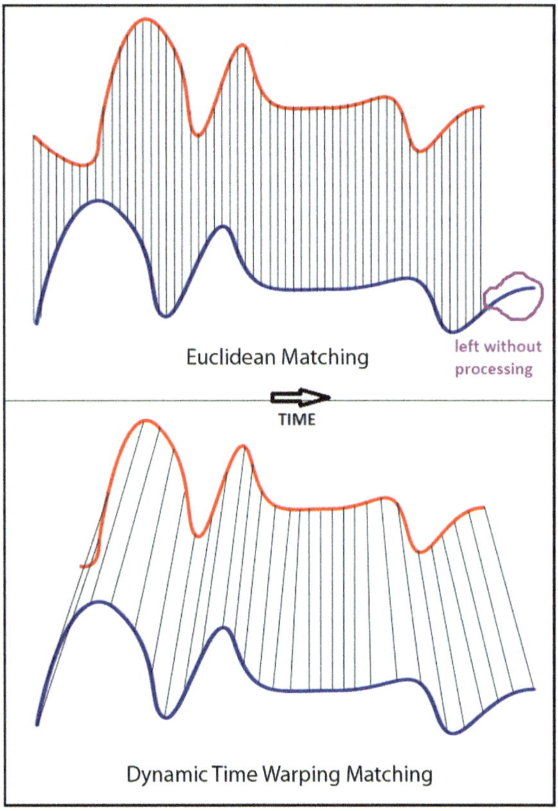

Fig. 3.3 The differences between Euclidean matching and DTW matching

and/or time axis distortion to estimate the data distances. This research uses a K-means clustering algorithm with a combination of dynamic time warping (DTW). As Fig. 3.3 shows, DTW has the capacity to establish a one-to-many match to ensure that troughs and peaks with the same pattern are precisely matched, and no curve is left without processing [4].

DTW has the capacity to be used for comparing arrays of varying lengths to create one-to-many, and many-to-one matches to minimize the overall distance [36]. Hence, it is used as a time alignment of time series to minimize the Euclidean distance between the aligned (i.e., resampled) time-series data where $\pi - [\pi_0, \ldots, \pi_K]$ is the approach to satisfy the following conditions:

1. a list of index pairs $\pi_k - (i_k, j_k)$ with $0 < i_k < n$ and $0 < j_k < m$
2. $\pi_0 - (0, 0)$and $\pi_k - (n - 1, m - 1)$
3. for all $k > 0$, $\pi_k - (i_k, j_k)$ is related to $\pi_{k-1} - (i_{k-1}, j - k - 1$ as follows:

   (a) $i_{k-1} \leq i_k \leq i_{k-1} + 1$
   (b) $j_{k-1} \leq j_k \leq j_{k-1} + 1$

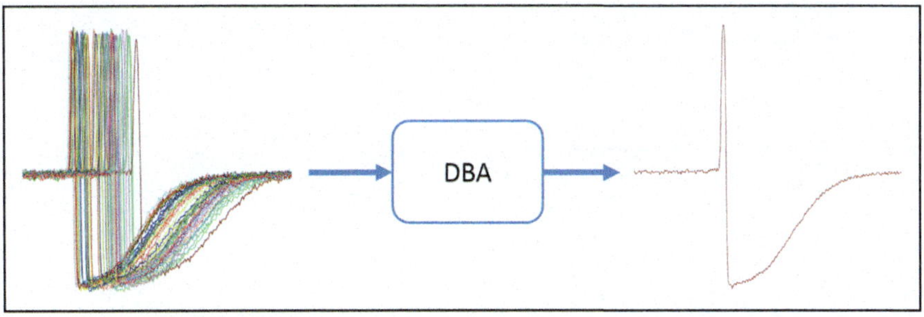

Fig. 3.4 DBA algorithm diagram interpretation

As Eq. 3.1 shows, DTW computes the square root of the sum of the distance (d) squares for each element in sequence X and the nearest point in sequence Y.

$$DTW(x, y) = \sqrt[\min_\pi]{\sum_{(i,j)\in\pi} d(x_i, y_j)^2} \tag{3.1}$$

K-means clustering takes the benefit of DTW to collect time series with similar shapes, and find the clustering centroid, named the gravity centre. The gravity center is the average sequence of a group of time series in DTW space. The DTW Barycenter Averaging (DBA) algorithm is used to minimize the sum of DTW distance squares between the barycenter and data sequence [24]. As Fig. 3.4 depicts, DBA returns an average shape of the cluster members regardless of the time shifts.

Silhouette score is used as the criteria for evaluating cluster performance. The score (from −1 to +1) represents the quality of clustering ranging from faulty to dense Özkoç [36]. It is calculated using Eq. 3.2.

$$S = \frac{b - a}{\max(a, b)} \tag{3.2}$$

3.3.3 *Predictive Analysis*

This section describes the process of Bitcoin price prediction. The methodology flowchart is depicted in Fig. 3.5.

Machine learning models, including Long-Short Term Memory (LSTM) [12] and Random Forest (RF) [8] are deployed using Elephas. Elephas [19] is an extension of the Keras Machine Learning framework. It has the capacity to build a distributed deep learning approach on Spark.

LSTM is an artificial Recurrent Neural Network (RNN) to address the conventional RNN's vanishing gradient problem. It includes three gates: an input gate that determines what should be saved, a forget gate that refers to what should be removed, and a sigmoid gate that focuses on how much data may flow through the gate. These

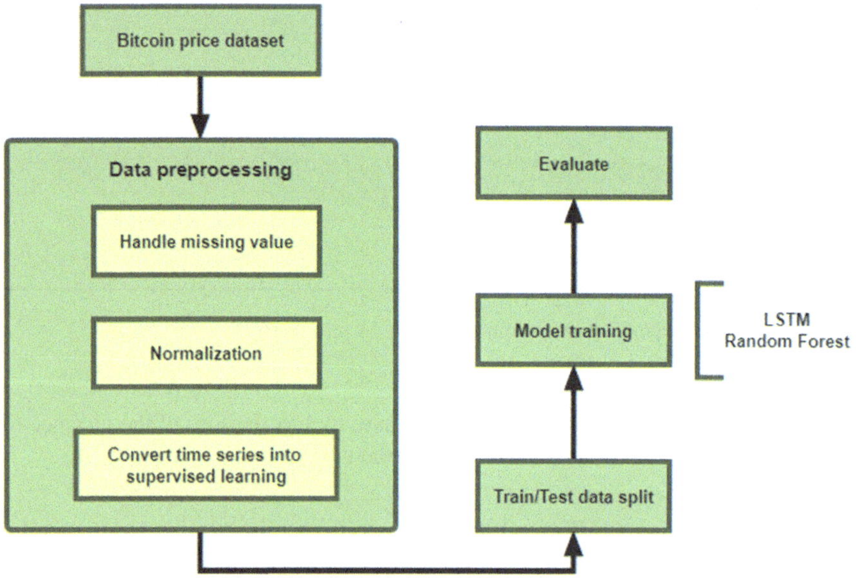

Fig. 3.5 Cryptocurrency price prediction flowchart

three gates are used to control data manipulations in the LSTM machine learning model, such as storing, writing, and retrieving data and predicting cryptocurrency price values.

Random forest is the collection of decision tree predictors for which the values of a random vector are independently and evenly sampled across all the trees. It has the capacity to outperform support vector machines and neural network classifiers in terms of robustness and over-fitting, especially for large has [17, 27]. Random forest is used to predict the market price trends (e.g., Bitcoin price) using the predictive data features data, including daily close price, open price, and Bitcoin volume.

To build and evaluate the models, the dataset is partitioned into two parts: training and testing. The ratio of the training data and the test data is 7:3. According to the dataset, it is assumed that the Bitcoin trend for each day will be the same as the previous records without the Fear&Greed Index (FGI) [1]. However, we include a feature called FGI in the model training to study the trend impact on cryptocurrency prices. The results are compared to whether FGI impacts the price predictions. By this, four different prediction scenarios are proposed as Table 3.1 shows.

Table 3.1 The experiment setup overview

	Features	Label
LSTM	Price of previous 30 min + FGI	Price value and trend
	Price of previous 30 min	Price value and trend
RF	Open, high, low, close, volume, FGI	Trend
	Open, high, low, close, volume	Trend

3.4 Result and Discussion

This section presents and discusses the experimental results of the data clustering and machine learning models.

The K-means clustering aims to recognise the data patterns. As Fig. 3.6 shows, the behaviour of the time-series data is represented by the grey lines, while the red lines represent the clustering centroid (the centre of gravity). By this, the dataset is divided into four partitions based on the similarities of the cryptocurrency prices. Principal Component Analysis (PCA) [15] is combined with the classical K-means algorithm to reduce the dimension of the original time-series data and improve the performance of the cluster and avoid overfitting problems.

3.4.1 *Trend Prediction*

The hit rate is used to evaluate trend prediction. The hit rate is the proportion of correct predictions over all the predictions. Table 3.2 shows the results of the hit rate. According to it, emotional factors result in a higher hit rate.

This study calculates profit to analyse the market trends. It aims to analyse Bitcoin trade profit according to the prediction results. According to this, people buy Bitcoins when the price drops and sell them when the price is increased. Figure 3.7 shows the daily Bitcoin profit.

Figure 3.8 shows the actual price and predicts Bitcoin price using LSTM. Although the curves have similar behaviours, the price prediction is still a regression problem. For this, four variables should be used to measure the results:

a. Mean absolute error (MAE): MAE measures the average of the residuals in the dataset. A smaller MAE refers to a better model. The calculating equation is $MAE = \frac{1}{N}\sum_{i=1}^{N}\left|y_i - \hat{y}\right|$, where $\hat{y}$ is the predicted value and y_i is the actual value for i-th sample.
b. Mean squared error (MSE): MSE measures the variance of the residuals. The calculating equation is $MSE = \frac{1}{N}\sum_{i=1}^{N}(y_i - \hat{y})^2$.
c. Root mean square error (RMSE): RMSE is the square root of the Mean Squared error. It measures the standard deviation of residuals.
d. R^2 score: R-squared represents the proportion of the variance in the dependent variable, which is explained by the linear regression model. The value of R2 should not exceed 1, and a larger R2 refers to a better model. A value equal to 1 shows an ideal model that makes no errors. However, a zero R2 means that the model performs as well as the baseline model, and a negative R2 means that the model does not perform as well as the baseline model.

Table 3.3 shows these four values for each experiment. It shows that the overall errors of LSTM are smaller than the benchmark.

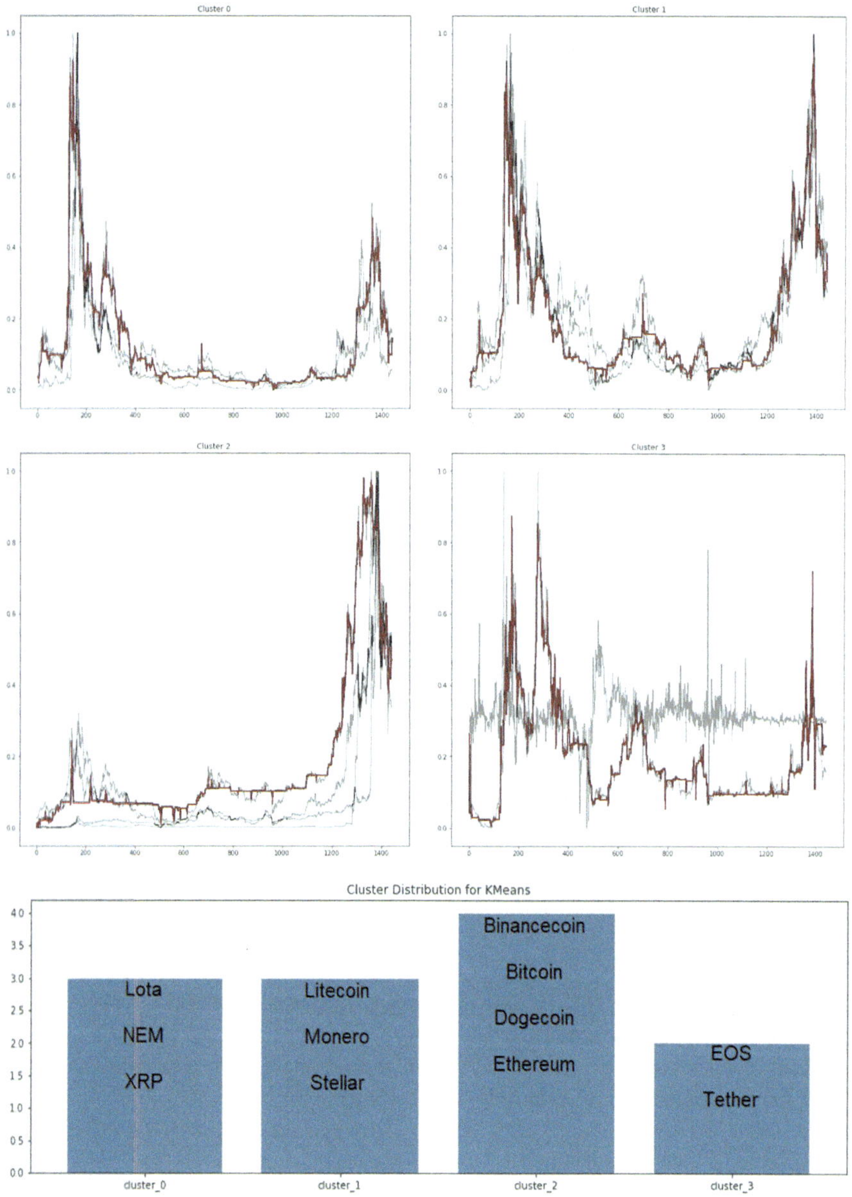

Fig. 3.6 Clustering results

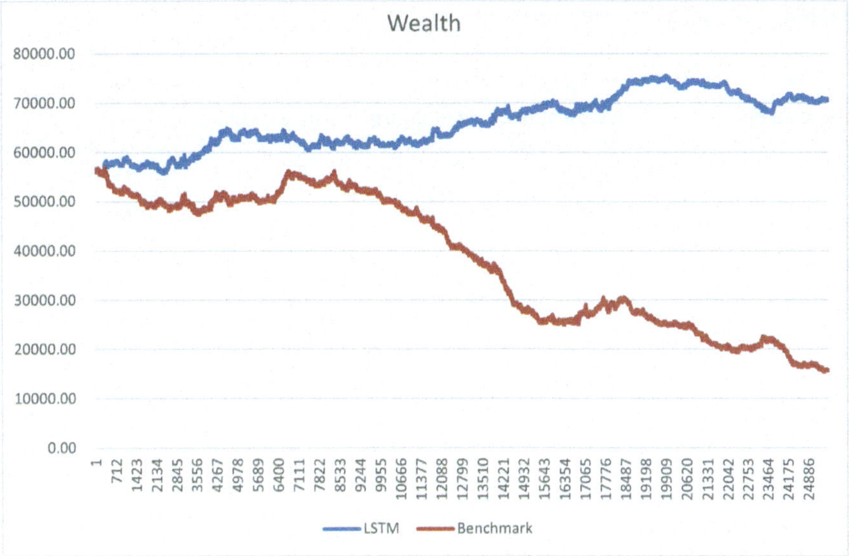

Fig. 3.7 The amount of wealth owned over time

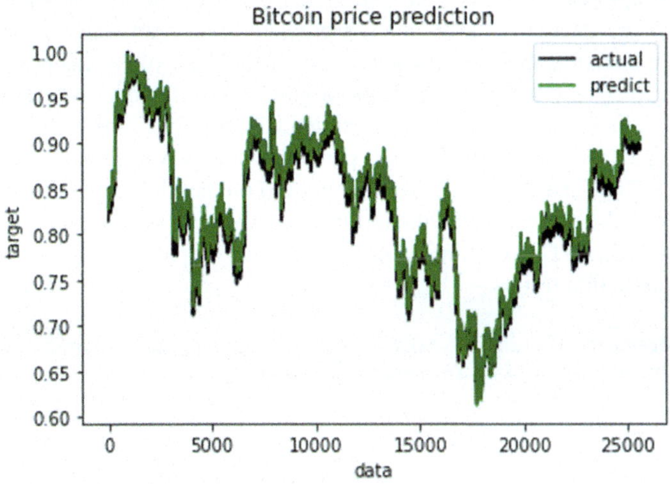

Fig. 3.8 The actual price and predict price using LSTM

Table 3.2 Hit rate for price trend prediction

	Case description	Hit rate
LSTM	Price of previous 30 min	0.498
Benchmark	Assume that the trend of the day will be the same as the previous day	0.480
RF	Open, high, low, close, ..., FGI	0.537
	Open, high, low, close, ...	0.504

Table 3.3 Results for price value prediction

	MSE	MAE	RMSE	R2
LSTM	0.00013	0.011	0.0113	0.978
Benchmark	4.57	0.0016	0.0021	0.999

3.5 Conclusion

This chapter proposes a Big data approach to study, analyse, model and predict cryptocurrency price patterns. It prepares and utilises a huge and variant time-series cryptocurrency dataset to train and test predictive machine learning models. For this, a normalization approach is used to rescale the data, yet, a data pre-processing approach is proposed to down-sample the original time-series data to decrease sample frequencies. In addition, a K-means clustering with dynamic time warping is used to recognise and study the cryptocurrency patterns, while two machine learning models including LSTM and Random Forest are used to predict the cryptocurrency values. According to the results, the cryptocurrencies have similar behaviours/trends in the market, and the LSTM machine learning technique outperforms Random Forest in terms of prediction results.

Acknowledgements The research work in this chapter was supported by our research team members: Sijin Wang, Shuguang Lyu, Ye Zhao, and Zhiyuan Lyu

References

1. Crypto fear and greed index—bitcoin sentiment. https://alternative.me/crypto/fear-and-greed-index/. Retrieved Nov 2022
2. Cryptocurrency historical prices. https://www.kaggle.com/datasets/sudalairajkumar/cryptocurrencypricehistory?select=coin_Aave.csv. Retrieved Mar 2023, July 2022
3. Mathew A (2020) Studying the patterns and long-run dynamics in cryptocurrency prices. J Corp Account Finan 31(3):98–113
4. Saeed A, Shirkhorshidi AS, Wah TY (2015) Time-series clustering—a decade review. Inform Syst 53:16–38

5. Baek U-J, Shin M-G, Lee M-S, Kim B, Park J-T, Kim M-S (2020) Comparison of distance measurement in time series clustering for predicting bitcoin prices. In: 2020 21st Asia-Pacific network operations and management symposium (APNOMS), pp 267–270. https://doi.org/10.23919/APNOMS50412.2020.9236969
6. Baig A, Blau BM, Sabah N (2019) Price clustering and sentiment in bitcoin. Finan Res Lett 29:111–116. ISSN 1544-6123. https://doi.org/10.1016/j.frl.2019.03.013
7. Barradas A, Tejeda-Gil A, Cantón-Croda R-M (2022) Real-time big data architecture for processing cryptocurrency and social media data: a clustering approach based on k-means. Algorithms 15(5). ISSN 1999-4893. https://doi.org/10.3390/a15050140
8. Leo B (2001) Random forests. Mach Learn 45(1):5–32
9. Jill C (2021) Cryptocurrencies: a guide to getting started. Technical report, World Economic Forum
10. Guo T, Bifet A, Antulov-Fantulin N (2018) Predicting short-term bitcoin price fluctuations from buy and sell orders
11. Hochreiter S, Schmidhuber J (1997) Long short-term memory. Neural Comput 9(8):1735–1780
12. Sepp H, Jürgen S (1997) Long short-term memory. Neural Comput 9(8):1735–1780
13. Hotz-Behofsits C, Huber F, Zorner TO (2018) Predicting crypto-currencies using sparse non-Gaussian state space models. Papers (5)
14. Jalali M, Heidari H (2020) Predicting changes in bitcoin price using grey system theory. Finan Innov 6
15. Ian TJ, Jorge C (2016) Principal component analysis: a review and recent developments. Philos Trans Royal Soc A Math Phys Eng Sci 374(2065):20150202. https://doi.org/10.1098/rsta.2015.0202
16. Kolla BP (2020) Predicting crypto currency prices using machine learning and deep learning techniques. Int J Adv Trends Comput Sci Eng 9(4)
17. Andy L, Matthew W et al (2002) Classification and regression by randomforest. R News 2(3):18–22
18. Maiti M, Vukovic D, Krakovich V, Pandey MK (2020) How integrated are cryptocurrencies. Int J Big Data Manage 1(1):64–80
19. Maxpumperla. elephas: distributed deep learning with keras and spark. https://github.com/maxpumperla/elephas. Retrieved Jan 2023
20. Mbanga Cedric L (2019) The day-of-the-week pattern of price clustering in bitcoin. Appl Econ Lett 26(10):807–811. https://doi.org/10.1080/13504851.2018.1497844
21. Monrat AA, Schelen O, Andersson K (2019) A survey of blockchain from the perspectives of applications, challenges, and opportunities. IEEE Access 7:117134–117151. https://doi.org/10.1109/access.2019.2936094
22. Naimy VY, Hayek MR (2018) Modelling and predicting the bitcoin volatility using Garch models. Int J Math Model Numer Optim 8(3):197
23. Ozer F, Okan Sakar C (2022) An automated cryptocurrency trading system based on the detection of unusual price movements with a time-series clustering-based approach. Exp Syst Appl 200:117017. ISSN 0957-4174. https://doi.org/10.1016/j.eswa.2022.117017
24. Paparrizos J, Gravano L (2015) k-shape: efficient and accurate clustering of time series. In: Proceedings of the 2015 ACM SIGMOD international conference on management of data, pp 1855–1870
25. Pele DT, Wesselhöfft N, Härdle WK, Kolossiatis M, Yatracos YG (2020) A statistical classification of cryptocurrencies. Available at SSRN 3548462
26. Phillips RC, Gorse D (2018) Cryptocurrency price drivers: wavelet coherence analysis revisited. PloS One 13(4):e0195200
27. Rodriguez-Galiano VF, Ghimire B, Rogan J, Chica-Olmo M, Rigol-Sanchez JP (2012) An assessment of the effectiveness of a random forest classifier for land-cover classification. ISPRS J Photogramm Rem Sens 67:93–104
28. Septiarini TW, Taufik MR, Afif M, Masyrifah AR (2020) A comparative study for bitcoin cryptocurrency forecasting in period 2017–2019. J Phys Conf Ser 1511(1):012056 (10pp)
29. Shah D, Kang Z (2014) Bayesian regression and bitcoin. IEEE

30. Shu M, Zhu W (2020) Real-time prediction of bitcoin bubble crashes. Physica A Stat Mech Appl 548:124477. ISSN 0378-4371. https://doi.org/10.1016/j.physa.2020.124477
31. Sigaki HYD, Perc M, Ribeiro HV (2019) Clustering patterns in efficiency and the coming-of-age of the cryptocurrency market. Sci Rep 9(1):1–9
32. Steinmetz F (2021) Behavioural clusters of cryptocurrency users: frequencies of non-speculative application domains. Technical report, BRL Working Paper Series
33. Wei Z, Pengfei W, Xiao L, Dehua S (2018) Some stylized facts of the cryptocurrency market. Appl Econ 50(55):5950–5965
34. Zhang Y, Ardakani SP, Han W (2021) Smart ledger: the blockchain-based accounting information recording protocol. J Corp Account Finan 32(4):147–157
35. Zielak. Bitcoin historical data. https://www.kaggle.com/datasets/mczielinski/bitcoin-historical-data. Retrieved Sept 2022, Apr 2021
36. Özkoç EE (2021) Clustering of time-series data. In: Birant D (ed) Data mining, chapter 6. IntechOpen, Rijeka. https://doi.org/10.5772/intechopen.84490

Chapter 4
Big Data Analytics for Credit Risk Prediction: Machine Learning Techniques and Data Processing Approaches

4.1 Introduction

Banks or financial institutions usually follow a review process of the "5C Principle", which are Capacity, Capital, Collateral, Conditions, and Character [23] to analyse their consumers' credits and outline the risks. This is used as a standard guideline to asses the eligibility of a customer to receive a loan or mortgage. This method works but each loan has to be manually inspected by a human being. This slows down the process and makes it tedious for both banks and individuals to reach an agreement.

Banking industries always strive to attain better profits. This can be achieved by maximizing the number of loans given, but it may be disastrous if the risk analysis is wrongly managed. A loan is very profitable for a bank, but it also carries the risk of falling out as some customers may not be able to pay back on time. Hence, it plays a key role to make fast and precise decisions which eliminate or minimise the loan and/or mortgage risks. Machine learning and Big data analysis solutions can be used to analyse customers' credit risk to make faster and more accurate financial decisions in banks or financial institutions.

This chapter aims to take the benefits of two well-known machine learning models including Decision Tree and Gradient Boosting to analyse and model credit risks and predict the behaviour of lenders. For this, a Pyspark framework is used to build and test the models, and a well-known and public dataset [21] is used to train the models. Moreover, a Big data-enabled data pre-processing approach is used to clean and prepare the dataset. The performance of both the models are measured, compared and reported in terms of Area Under ROC Curve (AUC), Area Under Precision-Recall curve (AUPRC), and model training time.

This chapter is organised as: Sect. 4.2 that reviews literature and introduces the relevant state-of-the-art solutions. Section 4.3 presents the methodology, while Sect. 4.4 discusses the experimental results. Section 4.5 concludes the chapter and addresses future works.

S. Pourroostaei Ardakani and A. Cheshmehzangi, *Big Data Analytics for Smart Urban Systems*, Urban Sustainability, https://doi.org/10.1007/978-981-99-5543-5_4

4.2 Literature Review

Banks, while highly profitable, are not invincible to financial problems. In some cases, a bank may go bankrupt due to poor management, risky endeavours, or mounting debt. According to a report by Federal Deposit Insurance Corporation (FDIC), in the last 5 years, there have been a total of 8 banks that have failed, with a total lost of 672 Million dollars [14]. Some of these banks have partially failed due to mounting debt caused by failed loans, as the bank has somehow lent more than they could. Banks everywhere are always wary of the possibility of failed loans. This problem leads to a standardization for all loaning activities that banks tend to follow.

Loan losses are a very significant event that every bank tries to avoid. Although many attempts are made to reduce these losses, it still happens. A recent quarterly review of banks has reported that banks has more than doubled its provisions.[1] Collectively, a sample of 70 international banks has set aside $161 Billion dollars in the first half of 2020, compared to the $50 billion in the second half of 2019 [11]. This huge increase in provisions was mostly caused by the global pandemic, which has affected multiple aspects of our daily lives. One of which is the rise of mortgages. In the first quarter of 2020, it was reported that the total debt reached a staggering amount of $10 Trillion dollars and is still increasing [15].

As stated above, it is critical for a bank to minimize the possibility of losing a loan. For this problem, machine learning models are created by researchers to predict bad loans. Sealand [25] concluded that the best performing algorithm was a mix of boosting classifiers and ensembled decision trees. Miao [22] has found that a sufficiently trained model will always outperform its human counterpart. An extra benefit is that a machine learning model keeps improving as more data is acquired.

This has helped financial institutes in ways that exceeded their original intention. Companies now have started to use machine learning models due to their ability to automate, ever-improving predictive capability, and its advanced data modelling [18]. For some countries, such as the United Kingdom (UK), it was found that the use of machine learning has risen quickly. From a surveyed 300 financial markets participants, 66% is now employing the use of machine learning [8]. The employment of machine learning in UK has improved operational efficiency by processing a greater volume of loan applications. The survey also reported that around 50% of banks in the UK has found machine learning to be important for future operations [8].

But not all machine learning models may be good enough to be employed for real-world use. Many methods have been used to find a model that is able to perform consistently. Each model has its own advantages and disadvantages that confounded researchers, as even with large amounts of data, most models have trouble with predictive accuracy. A study focusing on mortgage delinquency has found that a wide spread of model types are unable to consistently perform well enough to be marked as a solution to this problem [9]. This problem may occur due to the use of models that put strong assumptions on the linear relationship between output and

[1] A set amount of money set aside by a bank to cover future loan losses.

input variables. Some situations might have independent variables that are highly correlated to the output decision [26].

According to research focusing on the prediction of mortgage loans, not many exist that specifically target the current problem of predicting delinquency in mortgage loans [2]. Most studies and scholarly research work focus entirely on the end result of failed loans or Default loans [5], for example [6, 12, 13]. This chapter focuses on predicting the delinquency of mortgages as there is a much higher chance of a loan being labelled as Delinquency[2] and as a preemptive effort to avoid the default status altogether. However, mortgage loans rarely enter delinquency status. A statistic report [16] found that delinquency rates are decreasing every year in the United States (US). This is also one of the many reasons mortgage predictions pose a challenge to researchers. Many models are simply not "trained" enough due to the small amount of data that is labelled as delinquent. Though various modelling techniques and studies can be found that propose a method to improve performance, such as [3, 4, 7, 10, 19, 20], the same problem still persists. Some studies have tried to overcome the challenge by using Gradient Boosting.[3] One such study [1] has found that gradient boosting outperforms all other traditional models (such as Random Forest, Bagging, and Logistic Model Tree.) on most datasets.

4.3 Methodology

4.3.1 Dataset and Data Pre-processing

This study uses a Big and public dataset, named "Single Family Loan-Level", 2017–2019 [21]. Freddie Mac is the Federal Home Loan Mortgage Corporation, the US's major government-sponsored enterprise. A mortgage is a loan that is used to purchase real estate or to obtain funds by pledging an existing property. This dataset consists of two files: the mortgage origination data file, which indicates the information when the mortgage is initiated, and a monthly performance data file about the repayment performance of the associated loans and the delinquent status. The lender promises to repay the lender over a specified period of time, typically in a series of monthly installments divided into principal and interest. The property is pledged as collateral for the loan's repayment.

A data preprocessing approach is used to clean and prepare the dataset for machine learning model training [27]. The first step is to discard abnormal samples, missing values and outliers, and irrelevant/duplicate variables to achieve a high-quality dataset and avoid the expected impact on the prediction results. Then, the dataset is analysed to extract the meaningful features and figure out the data correlations. The key data features are outlined as below:

---

[2] A term for someone who has an overdue debt.

[3] A machine learning technique that builds a simple prediction model sequentially, where each model tries to predict the error left over by the previous model.

Table 4.1 Delinquent

Value	Counts
0	0.998913
1	0.001087

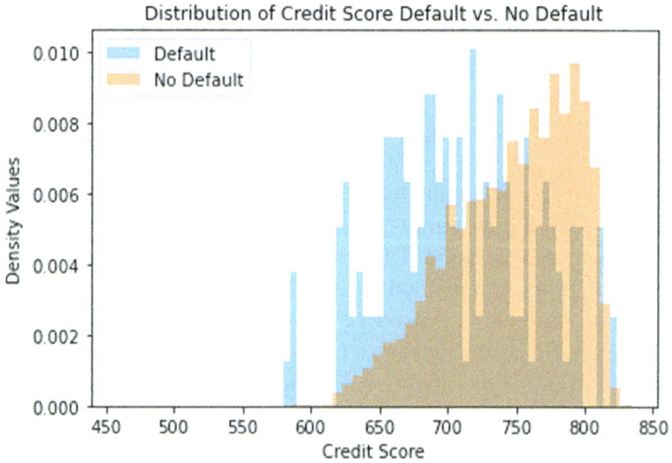

Fig. 4.1 Credit score after data clean

- **DELINQUENT**: the delinquency label of each loan is determined by leveraging the information available in **PREPAID** and **DELINQUENT**. If a loan is not delinquent, its value in DELINQUENT will be 0. If a loan is delinquent but also prepaid, its value in DELINQUENT will be 0 as well. Otherwise, the loan will be considered delinquent, and its DELINQUENT value will be 1. The value count result is shown in Table 4.1.
- **CREDIT SCORE**: according to [24], any credit rating system that enables the automatic assessment of the risk associated with a banking operation is called credit scoring. A higher credit score indicates that the borrower is relatively more able to repay the loan. For example, the likelihood of loan default is higher with a credit score of 600–700, while the likelihood of not defaulting on the loan is higher with a credit score of 700–800. Figure4.1 shows how the probability of loan default decreases if the credit score increases.
- **First Time Buyer Flag**: Table 4.2 gives the correlation of the delinquent rate and first-time buyers and shows that the first-time home-buyers are a weak predictor of loan delinquencies.

Table 4.2 Delinquent to first time buyer

Delinquent first time homebuyer flag	0	1
N	0.998924	0.001076
Y	0.998878	0.001122

Table 4.3 Delinquent to MSA

DELINQUENT metropolitan statistical area	0	1
48260.0	0.960000	0.040000
29700.0	0.964286	0.035714
48700.0	0.972222	0.027778
26300.0	0.978261	0.021739
29740.0	0.982759	0.017241

- **Metropolitan Statistical Area (MSA)**: Table 4.3 shows that lenders from all kinds of MSA give a similar probability of loan default.
- **Number of Units** and **Occupancy Status**: according to the dataset, 97% of the loans are for houses with just 1 unit and 88% of the home mortgages are primary residence, while *I* refers to investment property and *S* refers to a second home. Tables 4.4 and 4.5 shows that most primary residences investment properties and second homes are one-unit homes where 89% of one-unit homes are occupied by owners, and 81% of 4-unit homes are purchased as investment properties.
- **Combined Loan To Value**: Fig. 4.2 shows that the probability of a loan default is increased if the combined loan to value is enhanced.

  Table 4.6 shows that the loan default rate is almost twice as compared to when the loan to value is less than 90% where the combined loan to value is greater than 90%.
- **Debt to Income Ratio**: Fig. 4.3 shows that the higher debt-to-income ratio results in to a higher probability of a loan default.
- **Loan To Value**: this attribute refers to the amount of required loan for each lender and contains missing values similar to "Combined Loan To Value" feature. However, they are all replaced by a median value. Figure 4.4 shows a bimodal data distribution where peaks are at 70–80 and 90–100%. According to this, lenders require a loan amount of 75% (one the average) of the property's value. However, they tend to default more often if gravitate to the same loan or even more that the property's value.
- **Channel**: the majority of the loans are retail mortgage, which is a mortgage loan that is originated, underwritten, and funded by a lender or its affiliates. The followings are Correspondent and Broker.
- **Product Type**: this feature is discarded as it has the same FRM (Fixed Rate Mortgage) values.
- **Property Type**: as Fig. 4.5 shows, single-family home is largely the most popular property type among lenders. Table 4.7 shows the number of property units for each property type.
- **Number of Lenders**: Table 4.8 shows the density of the loans based on the number of lenders. According to this, most loans have only one lender.

Table 4.4 Occupancy statuses to number of units

Number of units occupancy status	1	2	3	4
I	0.844585	0.100524	0.025315	0.029576
P	0.987791	0.009623	0.001941	0.000645
S	1.000000	0.000000	0.000000	0.000000

Table 4.5 Number of units to occupancy statuses

Number of units occupancy status	1	2	3	4
I	0.070373	0.491587	0.546903	0.809417
P	0.889188	0.508413	0.453097	0.190583
S	0.040439	0.000000	0.000000	0.000000

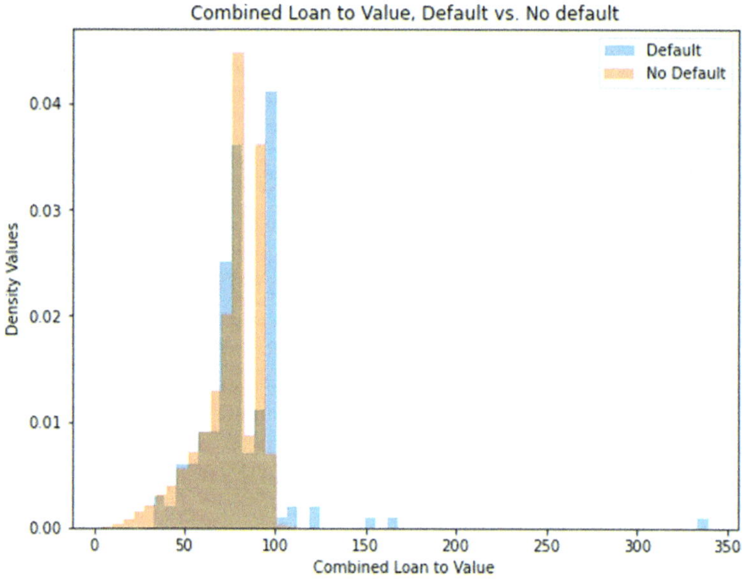

Fig. 4.2 Combined loan to value after data cleaning

Table 4.6 Combined loan to value

	> 90	≤ 90
0	0.998327	0.999059
1	0.001673	0.000941

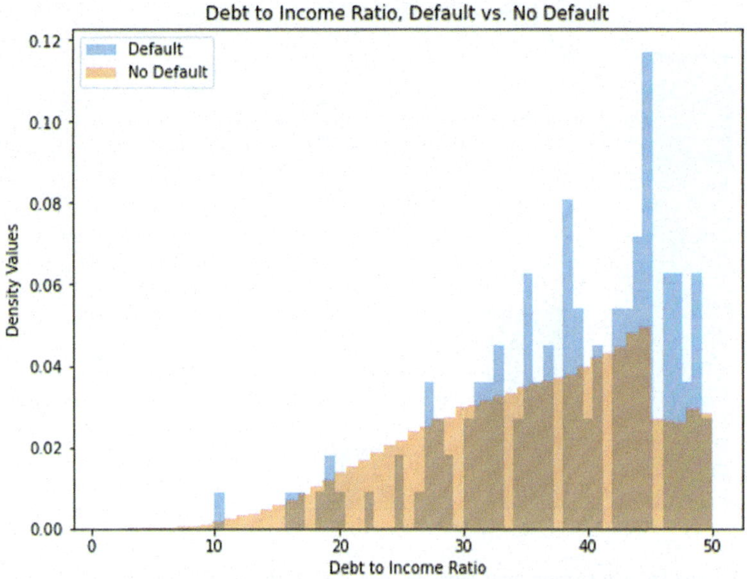

Fig. 4.3 Debt income to ratio after data cleaning

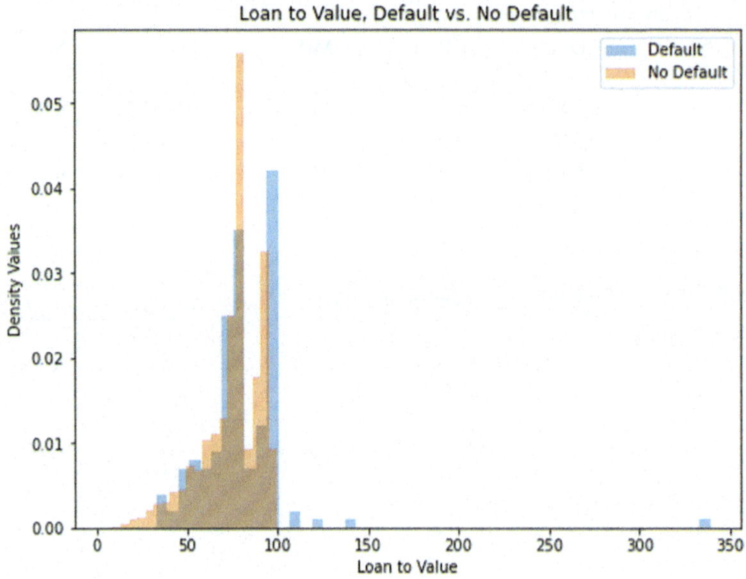

Fig. 4.4 Debt income to ratio after data cleaning

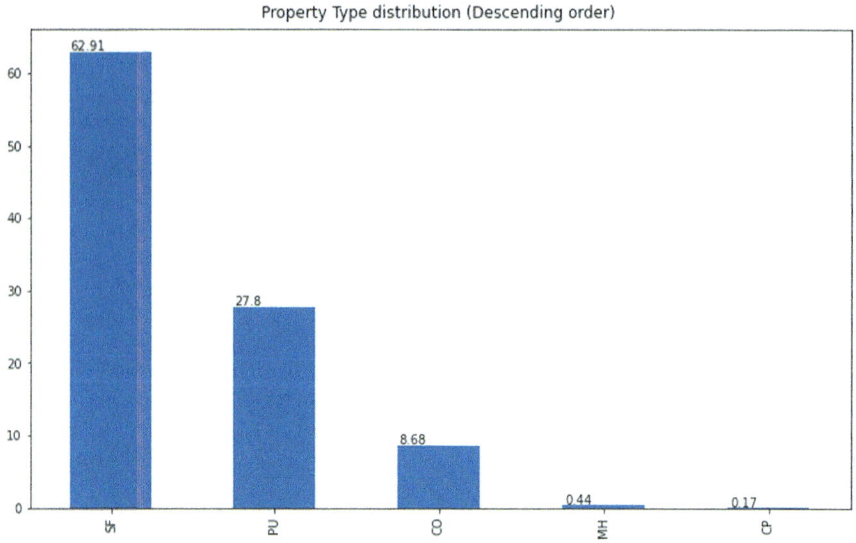

Fig. 4.5 Debt income to ratio after data cleaning. CO: Condo, PU: Planned Unit Development, MH: Manufactured Housing, SF: Single-Family, CP: Cooperative Share, and LH: Leasehold.

Table 4.7 Number of units and property types

Property type number of units	CO	CP	MH	PU	SF
1	0.999846	1.0	1.0	0.999161	0.963228
2	0.000154	0.0	0.0	0.000576	0.026175
3	0.000000	0.0	0.0	0.000048	0.005966
4	0.000000	0.0	0.0	0.000216	0.004631

Table 4.8 Number of lenders

Number of lenders	Density
1	0.533454
2	0.460553
3	0.005193
4	0.000780
5	0.000020

4.3.2 Machine Learning Models

A Pyspark framework is used to build and test Decision Tree (DT) and Gradient Boosting (GBDT) models. Decision Tree technique is chosen due to its simplicity and popularity in this field of research [7, 19], while Gradient Boosting Decision Trees (GBDT) are adaptable, easy to interpret, and give more accurate predictions [28]. The given dataset [21] contains 148,000 records. It is randomly partitioned as 70:30 ratio to form training and test datasets. Both models are trained to predict a single target as the status of delinquency. The performance of both the models are tested and evaluated according to Area Under ROC Curve (AUC), Area Under Precision-Recall curve (AUPRC), and model training time.

Boosting is a machine learning strategy to ensemble a series of weak predictive models (i.e., Decision trees) to improve learning. This extracts keys features from the original dataset to build the boosting tree [17].

$$\begin{aligned} F^* &= \arg\min_F E_{y,\mathbf{x}} L(y, F(\mathbf{x})) \\ &= \arg\min_F E_{\mathbf{x}} \left[E_y(L(y, F(\mathbf{x}))) \mid \mathbf{x} \right] \end{aligned} \tag{4.1}$$

Equation 4.1 presents the target generic function. By this, GBDT aims to measure an approximation $\widehat{F}(\mathbf{x})$ of function $F^*(\mathbf{x})$ mapping $\mathbf{x}$ to y, where the loss function L is minimal. The common loss function of Mean Squared Error (MSE) L is calculated using Eq. 4.2.

$$\begin{aligned} L_{\text{MSE}} &= \frac{1}{n}(y - F(x))^2 \\ \frac{\partial L_{\text{MSE}}}{\partial F} &= \frac{2}{n}(y - F(x)) = \frac{2}{n} h_m(x) \end{aligned} \tag{4.2}$$

4.4 Result and Discussion

The results of the two trained machine learning models are diverse. DT model is set-up with a Max depth of 10, Max bins of 10, and 42 seeds. As Table 4.9 reports, DT gives a prediction accuracy of 87.8%, however, it fails to properly identify and label Delinquent loans (AUC of 0.488) because the dataset is highly imbalanced. Indeed, DT gives a high accuracy and low AUC as the majority of correct perditions is from the Non-Delinquent labels.

The GBDT is built by `Spark, GBTClassifier` and uses both string index encoders and one-hot encoders for the categorical features. A greedy search technique is used to find the best-fitted model's hyper-parameters to tune the GBDT model. The model's hyper-parameters are listed in Table 4.10.

Table 4.11 reports the results and compares the performance of the both the proposed models in terms of AUC, AUPRC and model's training time. According to the

Table 4.9 Decision tree evaluation results

Amount of testing data	44,957 rows
Total correct prediction	39,436 (87.8%)
Actual delinquency count	5388
Predicted delinquency count	279
AUC (area under ROC)	0.488
AUPR (area under PR)	0.127

Table 4.10 GBDT's hyper-parameters

Parameter	Value
stepSize	0.01
seed	42
maxIter	20
maxDepth	10

Table 4.11 DT versus GBDT

	AUC	AUPRC	Training time (s)
Decision tree	0.488	0.127	50.67
Gradient boosting	0.987	0.968	41.43

results, GBDT outperforms DT and provides a better prediction result of delinquent mortgages. Moreover, it is able to reduce model training delay as compared to DT. This supports that boosting technique improves the performance of weak learners (i.e., DT) and increases perdition accuracy, especially in large dataset where features are highly imbalanced.

4.5 Conclusion

The purpose of this study was to take the benefits of machine learning and Big data analysis approaches to analyse credit risks and predict delinquent mortgages. The dataset was imbalance and needed a substantial pre-processing to eliminate abnormal samples, missing values and outliers, and irrelevant/duplicate features, and normalise the values. The prepared dataset was used by two machine learning models including DT and GBDT to analyse the potential credit risks for mortgages. As the results report, GBDT outperforms DT and gives better prediction results in terms of AUC, AUPRC, and model training time. However, utilising other machine learning techniques such as Support Vector Machine (SVM) to analyse credit risks and build predictive models can be considered as a future work.

Acknowledgements The research work in this chapter was supported by our research team members: Yizirui FANG, Yukai LU, Quanchi CHEN, and Ivan Christian HALIM.

References

1. Alam TM, Shaukat K, Hameed IA, Luo S, Sarwar MU, Shabbir S, Li J, Khushi M (2020) An investigation of credit card default prediction in the imbalanced datasets. IEEE Access 8:201173–201198. https://doi.org/10.1109/ACCESS.2020.3033784
2. Azhar Ali SE, Rizvi SSH, Lai F, Faizan Ali R, Ali Jan A (2021) Predicting delinquency on mortgage loans: an exhaustive parametric comparison of machine learning techniques. Int J Ind Eng Manage 12(1):1–13
3. Baesens B, Van Gestel T, Viaene S, Stepanova M, Suykens J, Vanthienen J (2003) Benchmarking state-of-the-art classification algorithms for credit scoring. J Oper Res Soc 54(6):627–635
4. Bellotti T, Crook J (2009) Credit scoring with macroeconomic variables using survival analysis. J Oper Res Soc 60(12):1699–1707
5. Berson K (2023) 4 things to know about defaulting on your mortgage. https://upsolve.org/learn/mortgage-default/. Retrieved Feb 2023
6. Bracke P, Datta A, Jung C, Sen S (2019) Machine learning explainability in finance: an application to default risk analysis
7. Brown I, Mues C (2012) An experimental comparison of classification algorithms for imbalanced credit scoring data sets. Expert Syst Appl 39(3):3446–3453
8. Buchanan BG, Wright D (2021) The impact of machine learning on UK financial services. Oxford Rev Econ Policy 37(3):537–563. https://doi.org/10.1093/oxrep/grab016
9. Chen S, Guo Z, Zhao X (2021) Predicting mortgage early delinquency with machine learning methods. Euro J Oper Res 290(1):358–372
10. Crook JN, Edelman DB, Thomas LC (2007) Recent developments in consumer credit risk assessment. Euro J Oper Res 183(3):1447–1465
11. de Araujo DKG, Cohen BH, Pogliani P (2023) Bank loan loss provisioning during the covid crisis. https://www.bis.org/publ/qtrpdf/r_qt2103w.htm. Retrieved Mar 2023
12. de Castro Vieira JR, Barboza F, Sobreiro VA, Kimura H (2019) Machine learning models for credit analysis improvements: predicting low-income families' default. Appl Soft Comput 83:105640
13. Deng Y, Quigley JM, Order RV (2000) Mortgage terminations, heterogeneity and the exercise of mortgage options. Econometrica 68(2):275–307
14. FDIC (2023) Bank failures in brief—summary 2001 through 2022. https://www.fdic.gov/bank/historical/bank/. Retrieved Mar 2023
15. Fontinelle A (2023) American debt: mortgage debt reaches $10.04 trillion in q4 2020. https://www.investopedia.com/personal-finance/american-debt-mortgage-debt. Retrieved Apr 2023
16. FRED Economic Data. Delinquency rate on single-family residential mortgages, booked in domestic offices, all commercial banks. https://fred.stlouisfed.org/series/DRSFRMACBS. Retrieved Mar 2023
17. Friedman JH (2001) Greedy function approximation: a gradient boosting machine. Annal Stat: 1189–1232
18. Imran M (2023) How machine learning is being used by mortgage company. https://www.folio3.ai/blog/how-machine-learning-is-being-used-by-mortgage-company/. Retrieved Mar 2023
19. Kennedy K, Namee BM, Delany SJ (2013) Using semi-supervised classifiers for credit scoring. J Oper Res Soc 64(4):513–529
20. Lessmann S, Baesens B, Seow H-V, Thomas LC (2015) Benchmarking state-of-the-art classification algorithms for credit scoring: an update of research. Euro J Oper Res 247(1):124–136

21. Mac F (2019) Single family loan-level dataset. Available online: http://www.freddiemac.com/research/datasets/sf_loanlevel_dataset. Accessed on 13 Apr 2022
22. Miao L (2022) Assessing the strengths and weaknesses of human information processing in lending decisions: a machine learning approach. J Account Res 60(2):607–651
23. Navy Federal (2023) Understanding the "Five C's" of credit. https://www.navyfederal.org/resources/articles/small-business/the-5-cs-of-credit.html. Retrieved Mar 2023
24. Pérez-Martín A, Pérez-Torregrosa A, Vaca M (2018) Big data techniques to measure credit banking risk in home equity loans. 89:448–454
25. Sealand JC (2018) Short-term prediction of mortgage default using ensembled machine learning models. Master's thesis, School of Mathematics and Statistics, Slippery Rock University
26. Wager S, Athey S (2018) Estimation and inference of heterogeneous treatment effects using random forests. J Am Stat Assoc 113(523):1228–1242
27. Wang Q, Rajakani K (2022) Research on the method of predicting consumer financial loan default based on the big data model
28. Ye J, Chow J-H, Chen J, Zheng Z (2009) Stochastic gradient boosted distributed decision trees. https://doi.org/10.1145/1645953.1646301

Chapter 5
Worldwide Mobility Trends and the COVID-19 Pandemic: A Federated Regression Analysis During the pandemic's Early Stage

5.1 Introduction

Since the emergence of COVID-19 (Novel Coronavirus Disease 2019) in December 2019, the world has experienced a rapidly increasing number of infections daily [9, 34]. This large number of infected cases highlights the COVID-19 pandemic as a severe issue that needs an urgent and efficient solution. Humans are the target hosts of the virus, and limiting human contact can reduce the rate of virus spread and ease the outbreak [2]. As governments need to develop specific measures to control the epidemic quickly, they should find efficient solutions to manage travel trends and restrict people gathering to minimise close contact. In order to capture how people's travel behaviour is affected by the COVID-19 pandemic and how it changes over time, big data analysis offers a number of benefits -mainly predicting mobility trends to provide valuable insights rapidly and accurately for early outbreak prevention and control decisions.

People's mobility can be influenced by various parameters such as weather conditions, social events, or pandemics (i.e., COVID-19). During the COVID-19 outbreak, countries used various lockdown measures [21, 25] to control and prevent the pandemic resulting in people's mobility restrictions.

This chapter studies the sudden impacts of the COVID-19 pandemic on mobility trends at the early stage of the outbreak. It analyses people's mobility for retail and recreation, groceries and pharmacies, parks, transit stations, workplaces, and residential once the world copped with the virus spreading from February to April 2020. This research aims to model people's mobility trends at the early stage of a pandemic to manage resources and reduce potential risks. For this, an online and validated dataset from [6] is pre-processed and prepared to analyse people's mobility trends in 18 countries worldwide. The correlations between COVID-19 and mobility trends are highlighted via a correlation matrix technique. The most influential data features are extracted to train a regression model to forecast people's mobility trends. The outcomes of this research contribute to answering the following questions:

S. Pourroostaei Ardakani and A. Cheshmehzangi, *Big Data Analytics for Smart Urban Systems*, Urban Sustainability, https://doi.org/10.1007/978-981-99-5543-5_5

(1) How the COVID-19 outbreak has suddenly influenced community mobility trends?
(2) How the mobility trends were changed in different countries once the pandemic began?
(3) How were different sectors influenced by mobility trends during the COVID-19 pandemic?

The rest of this chapter is organised as follows: Sect. 5.2 reviews the literature and highlights the key factors influencing mobility trends. Section 5.3 presents the research methodology and highlights the features of the data analysis approach. Section 5.4 demonstrates data analysis results and describes how the COVID-19 epidemic influences mobility trends. Section 5.5 concludes the key findings of this research and addresses the future works..

5.2 Literature Review on Existing Research Studies

5.2.1 Influence Factors

Several factors influence the mobility of people, such as gender [16], age and personal choice [32], lifestyle habits [28], and occupation [8]. Nevertheless, additional factors such as population characteristics and different urban forms also significantly impact mobility trends [12, 13]. This varies significantly between the two different urban forms of polycentric and monocentric [30]. Furthermore, different cultural forms also implicitly influence the direction of change in the movement of people [12, 14]. With the outbreak of sudden epidemics, a new aspect has been added to the factors, and it became the main impact of people's daily travel.

5.2.2 Pharmacological and Non-pharmacological Interventions

The speed and severity of the COVID-19 outbreak are a great challenge to healthcare resources that are highly restricted. This addresses the inability to meet the demand for emergency care for infected patients during the epidemic [10, 29]. This means hospitals are unable to plan healthcare resources carefully in the case of a sudden pandemic. It can lead to a situation where the demand for intensive healthcare services such as ICU staff and equipment is significantly increased. This may result in insufficient healthcare services that leave many infected patients without treatment and increase infected and/or death cases. In the early stages of the epidemic, there was no in-depth knowledge of the disease and an effective vaccine to protect people from infection. In a short period of time, the shortage of medical resources became one of the factors for the rapid outbreak of the epidemic [7]. However, people could take benefit from a strong relationship between virus transmission and people mobility using the SIS model in epidemic networks to manage the pandemic [18]. Restricting the movement of people under conditions of non-pharmacological intervention is, therefore, one of the effective measures used by the government to control the spread of epidemics. For this, understanding people's mobility pattern during epidemics is important to provide sufficient information needed to formulate epidemic policy [11]. This addresses a data-driven analysis to help governors to explore mobility trends and propose effective and efficient solutions to minimise them.

5.2.3 *Social Distance Policy*

Wuhan lockdown was a very successful solution that led other countries to similarly attempt to control the movement of people across regions [24]. For example, mobility data collected from people's mobile phones support the mobility restrictions and shows that short- and long-term mobility has significantly fallen in France during the pandemic [26]. However, each government should accurately propose mobility limitation protocols to reduce the virus spreading with minimised impact on people's daily life. The epidemic's impact on mobility patterns is highly related to mobility infrastructures such as roads and rails. For this, mobility patterns and available infrastructure should be carefully studied to propose population mobility restriction policies and design better solutions to manage the epidemic. According to [15], the UK and France could better manage the pandemic as they have a centralized mobility network, whereas Italy was not very successful due to the lack of concentrated mobility infrastructure.

5.2.4 *Reflection of H1N1*

Mobility restrictions and social distance policies are the critical solutions that offer benefits to manage similar pandemics such as H1N1. Although (A)H1N1 had different clinical symptoms, it caused widespread infections when it started in Mexico [19]. To manage H1N1, social distancing and mobility restrictions could successfully manage the situation. Some countries have adopted control measures to restrict people's travel as part of their emergency policies in response to H1N1. This aimed to reduce the number of travel from infectious areas and limit contact with infected people. As a result, Mexico received a 40% reduction in international air traffic [22]. However, this was not enough to manage the pandemic. For this, the Global Epidemic and Mobility model (GLEaM) was used to simulate the movement of people during the H1N1 pandemic. GLEaM addresses an exponential growth dominating an area where an infected case is found. This result suggests that travel restrictions have a low impact on curbing the spread of the epidemic. This illustrates that the epidemic origin is characterized by spatial heterogeneity and intra-region mobility [4], meaning epidemics in different spaces have different characteristics. The factors that influence its spread within a region are also different. Therefore, more factors need to be taken into account when exploring the relationship between epidemic spread and human mobility.

5.2.5 Cultural Susceptibility and Policy

Travel restrictions are particularly useful in the early stages of an outbreak. However, this may become less effective if the outbreak becomes more widespread [20]. Thus, under a COVID-19 pandemic, it is not sufficient to impose travel restrictions similar to those in place during the H1N1 period. To control the rate of growth of the epidemic within a region, it is necessary to develop a policy of social distancing in each region. The analysis of intra-regional movements is an effective action for developing measures. Cultural susceptibility becomes an important factor when considering population movement patterns within a region. Cultural differences between regions can contribute to differences in public awareness and are reflected in the social distancing policies developed in each location. During the COVID-19 epidemic, in order to keep peak infection levels below the resource capacity of the healthcare system, a series of social distancing policies such as home orders, restrictions on restaurants and bars, and school closures were developed and imposed in the United States [1]. Compared to normal footfall, people's home activity increased during the epidemic, while other mobility in other locations decreased.

Lenient social-distancing policies have had a limited effect on reducing social interaction, and much of the social-distancing capacity of such policies has been absorbed by non-policy-driven mechanisms, with voluntary mechanisms being an essential factor in reducing mobility. In Wuhan, China, the government has taken unprecedentedly strict intervention measures. Suspected and confirmed cases and contacts are immediately quarantined and placed under medical observation. Moreover, the government has imposed traffic restrictions on Wuhan. The ban eliminated the possibility of access for most of the population. The interventions implemented in China have apparently been successful in reducing localised transmission and slowing the spread of COVID-19 [20]. It is worth noting that prior to the issuance of the Wuhan traffic restrictions, there was an increase in individuals leaving Wuhan, with the number higher than the same level in previous years. After the ban was issued, this number quickly dropped to nearly nothing. The intervention had a significant limiting effect in restricting population movements in and out of Wuhan. Comparing the epidemic data, it is clear that the more stringent interventions capture more of the positive impact. The analysis of changes in population movements by region should also take into account regional and ideological differences, which can lead to differences in the understanding of social distance policies and the extent to which the population implements them in each region.

5.2.6 Voluntary Mechanisms

Voluntary mechanisms were the main drivers of reduced human mobility in the early years of the epidemic, and social distancing behaviours undertaken voluntarily are evident in trends in geographical mobility [1]. The results of a traffic survey published by [33] show that the public transport system is the most affected in this respect, with a large part of the population refusing to use it in order to reduce the risk of transmission and to reduce social interaction. In areas where travel is restricted, the percentage decline in the population choosing public transport is always higher than those choosing private transport. Local medical organizations recommend avoiding public transport as much as possible and promoting the use of personal transport, such as bicycles, to get around. The population has voluntarily engaged in social segregation behavior, which is why people refuse to use the public transport system. The study results showed similar reductions in private vehicle and pedestrian flows, but to a lesser extent than for public transport services. The magnitude of the impact of voluntary mechanisms on population movements depends mainly on the population's awareness of the modes of transmission, infectiousness, and severity of COVID-19. The population generally has a good understanding of the main modes of transmission and common symptoms of COVID-19. However, a large proportion of the population has misconceptions about what can be done to prevent infection. An online survey revealed that many people believe wearing a general surgical mask is "very effective" in protecting them from COVID-19 [17]. This misconception stems from the limited access to information about the epidemic and the lack of awareness among healthcare providers. When people misjudge the severity of COVID-19, the social distancing behavior resulting from voluntary mechanisms will be reduced, and the restriction of population movement will be less than expected.

The impact of age on trip behavior can also be reflected in people's voluntary mechanisms. A study analyzed the intention of each age group for all trip combinations and long-distance trips larger than 100 km. The decrease in the total number of trips was evenly distributed across each age group, but for long-distance trips, the reductions increased with age [27]. Among those infected with COVID-19, the rate of severe disease is higher in older people [23], who might have exhibited increased risk aversion. People aged 24–59 are the most active, and usually, this group will congregate in an area for reasons such as work commuting, resulting in very high mobility in the area. When an outbreak occurs, some people will still leave their homes as often as they work, while others will stay home. Therefore, such areas are usually the areas where quarantine measures have the most significant impact on the mobility of people. Taking age characteristics into account as a factor when considering how to develop social distance measures will allow for more effective implementation.

Table 5.1 Data features (*source* The Authors)

Mobility trend sectors	COVID-19 data
Retail and recreation	Confirmed
Grocery and pharmacy	Recovered
Parks (leisure)	Death
Transit stations	World confirmed
Workplaces	World recovered
Residences	World death

5.3 Methodology

5.3.1 Data Sources

This study uses a public and clean dataset that includes daily community mobility in 120+ countries for six different sectors during the early stage of the COVID-19 pandemic from February to April 2020 (COVID-19 Community Mobility Reports 2021). Besides, this dataset provides COVID-19 data, including local and worldwide numbers of confirmed, recovered, and death cases. Table 5.1 outlines the data features in the dataset:

5.3.2 Statistical Analysis

This study trains a regression model to predict people's mobility trends according to the pandemic outbreak. It uses a correlation matrix method to analyse the data features and select the predictor and prediction features for the regression model. Figure 5.1 demonstrates the critical steps of the data analysis in this research.

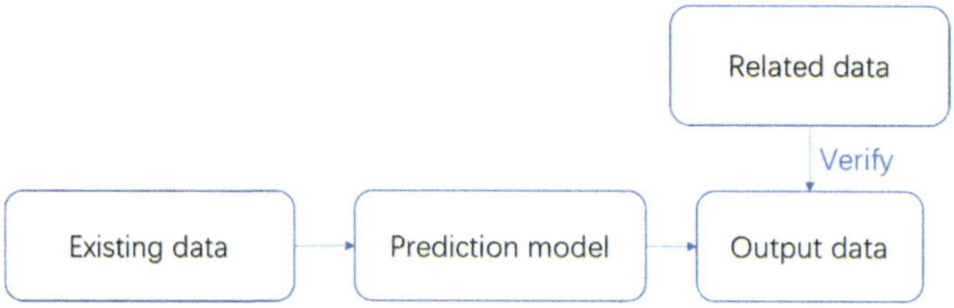

Fig. 5.1 The key steps of data analysis (*source* The Authors)

5.3.3 *Data Analysis*

A divide and conquer technique is used to investigate the impact of the COVID-19 outbreak on mobility trends and predict travel trends [5]. It recursively partitions the dataset into countries with similar situations and variation tendencies and trains local regression models. Then, the sub-models are aggregated to form the master model, which focuses on the whole dataset.

5.3.4 *Correlation Matrix*

The correlation matrix technique is used to investigate the relationship between people's mobility trends and COVID-19 [3]. The results form a table containing the correlation coefficients between each variable and the others (STHDA, n.d.) The correlation values demonstrate the dependency on community mobility and COVID-19 in each country.

5.3.5 *Regression*

The regression model predicts the mobility trends in each country according to the COVID-19 data. First, the correlation matrix highlights highly correlated features and extracts predictors and predictions. A prediction model is proposed to take the predictors and return prediction results for each region. Finally, a final model will be generated and tested by aggregating the local models and datasets.

5.4 Results and Discussion

5.4.1 *Correlation*

The correlation matrix helps determine the correlations between community mobility and pandemic outbreak. As Fig. 5.2 shows, 1 and -1 correlation values represent a positive and negative relationship between the two attributes. According to the results, community mobility correlates more with worldwide COVID-19 data than local COVID-19 in each country. According to the correction matrix, the infection period (number of days) highly influences mobility trends. It stems from the fact that the number of confirmed COVID-19 cases exponentially increased during each country's beginning of the pandemic. Besides, the correlation matrix shows that mobility trends in the five sectors, including retail and recreation, transportation, workplaces, grocery and pharmacy, and residential, have been highly influenced (above 0.8) by the outbreak. However, it shows that the impact of the pandemic on people's mobility trends to parks was the minimum for both national and worldwide levels.

Figure 5.3 depicts the mobility trends in five benchmark countries, including Italy, Australia, the USA, India, and Japan. According to this, Japan has the minimum mobility, while India and Italy show the maximum mobility trends compared to others. Besides, it shows that mobility trends to visit grocery, pharmacies, and parks differ from other sectors in these five countries.

Figure 5.4 shows time-series mobility trends for the six sectors. As it shows, there is an upward trend to stay home and visit residential sectors. However, people's mobility trends are sharply reduced to visit outdoor locations, even pharmacies and workplaces, during the outbreak.

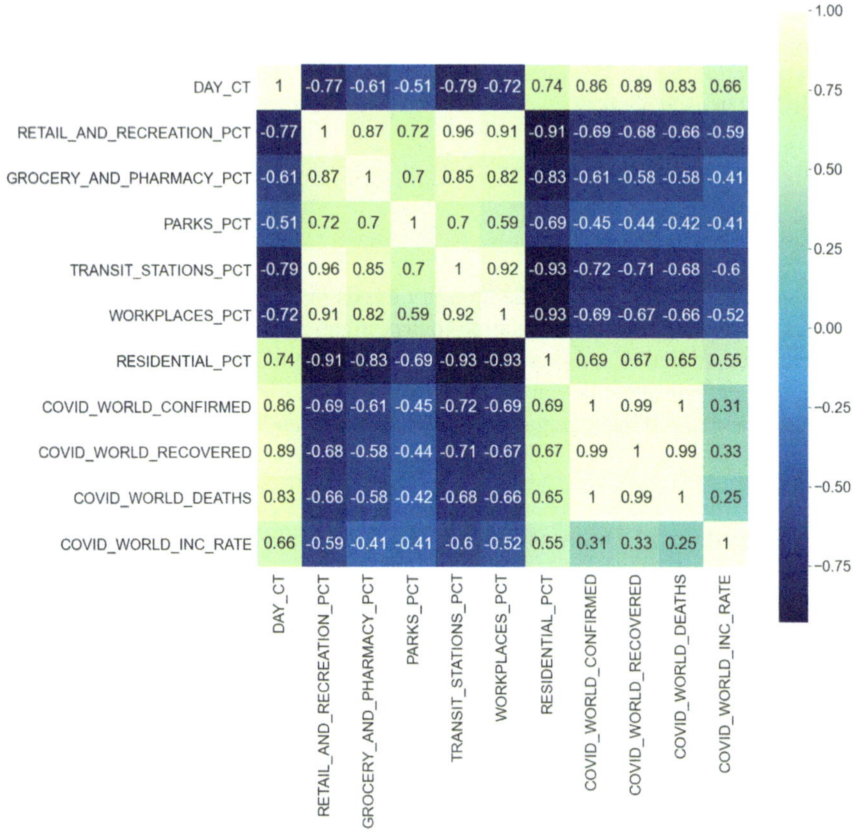

Fig. 5.2 Correlation of mobility trends and COVID-19 data (*source* The Authors)

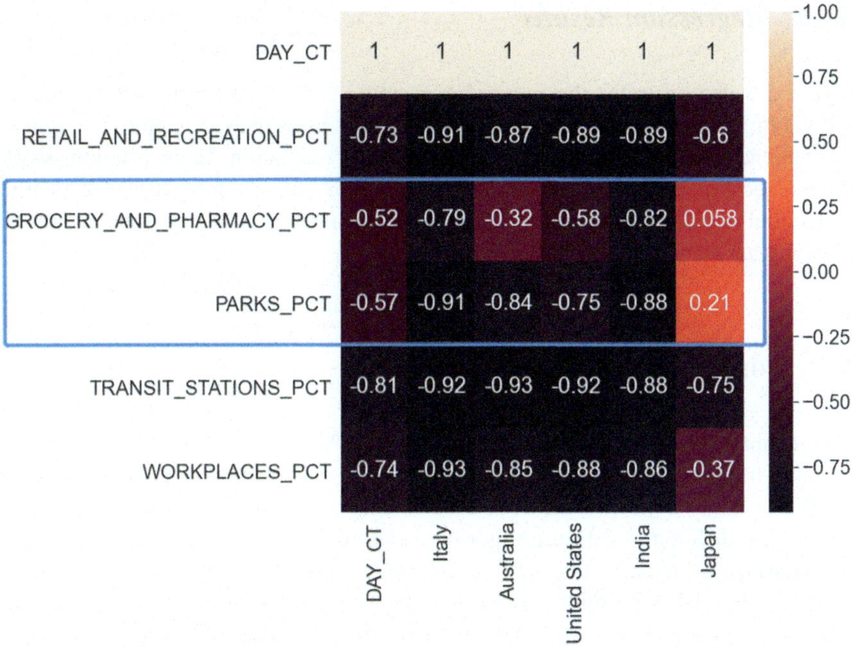

Fig. 5.3 Mobility trends for five benchmark countries (*source* The Authors)

Fig. 5.4 Worldwide mobility trends based on six studied sectors (*source* The Authors)

5.4.2 Regression Results

Figure 5.5 demonstrates the regression results of the mobility trends for the six sectors in 18 countries. As it shows, people's mobilities in these countries are more convergence and intend to decline for retail and recreation, transportation, workplaces, grocery, and pharmacy. Besides, they show an increasing trend of visiting residential sectors and homestays during the pandemic. However, these countries demonstrate different trends in park sectors.

5.5 Conclusions

This chapter aims to explore the impact of COVID-19 on people's community mobility trends at the early stage of the COVID-19 pandemic from February to April 2020. For this, an online dataset is used to analyse people's mobility in 18 countries to six different location categories, including retail and recreation, groceries and pharmacies, parks, transit stations, workplaces, and residential are analysed. A regression model is presented to predict the mobility trends in different countries if a similar pandemic appears in the future. This prediction will help governments know the expected mobility trend to regulate related policies that slow the epidemic's spread. In addition, this allows industry and business sectors to propose a predictive plan to maximised people's satisfaction.

This chapter takes the benefits of the correlation matrix and regression methods. The correlation matrix highlights the correlation between COVID-19 trends and mobility trends and outlines the factors which have the highest correlations and address a high impact. The results support that world-level COVID-19 factors (e.g., confirmed cases) highly correlate with mobility trends. Furthermore, the study highlights how early mobility restrictions could reduce the spread of the disease and have less longer-term impacts on the globe. There are still limitations in this study that should be addressed in the future. The study could only focus on a few countries rather than the global data. A much larger dataset would help us consider multiple aspects and variables to evaluate the trends. We also suggest that future research could consider scenario development as part of comparative analysis, for instance, in regards to how lockdown or other measures could help control or mitigate the disease spread at large.

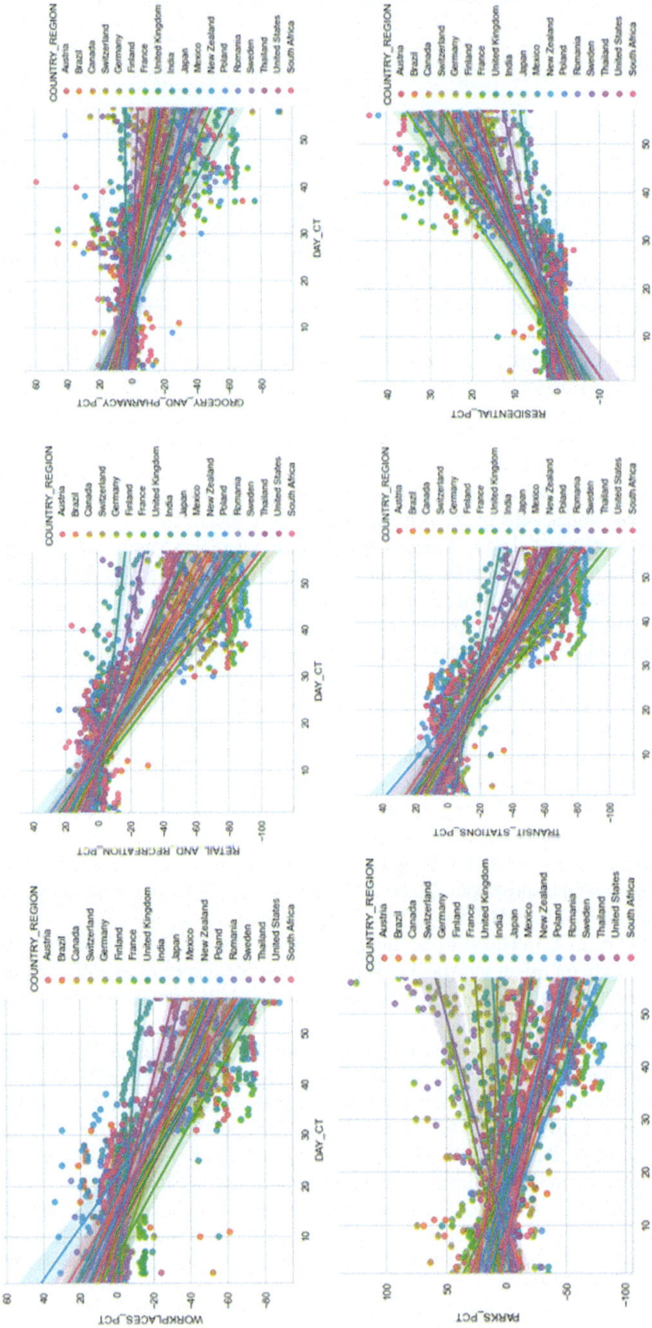

Fig. 5.5 Worldwide mobility trends based on selcted studied countries (*source* The Authors)

Acknowledgements The research work in this chapter was supported by our research team members: Qiyuan Zeng, Zhengduo Xiao, Yiming Li, and Shihong Huang.

References

1. Abouk R, Heydari B (2021) The immediate effect of COVID-19 policies on social-distancing behavior in the United States. Publ Health Rep 136(2):245–252
2. Alqithami S (2021) A generic encapsulation to unravel social spreading of a pandemic: an underlying architecture. Computers 10(1):12
3. Asuero AG, Sayago A, Gonzalez AG (2006) The correlation coefficient: an overview. Crit Rev Anal Chem 36(1):41–59
4. Bajardi P, Poletto C, Ramasco JJ, Tizzoni M, Colizza V, Vespignani A (2011) Human mobility networks, travel restrictions, and the global spread of 2009 H1N1 pandemic. PLoS ONE 6(1):e16591
5. Bentley JL, Haken D, Saxe JB (1980) A general method for solving divide-and-conquer recurrences. ACM SIGACT News 12(3):36–44
6. Chaudhury A (2021) COVID-19 community mobility dataset. Kaggle. Available from https://www.kaggle.com/arghadeep/covid19-community-mobility-dataset
7. Chen X, Liu Q, Wang R, Li Q, Wang W (2020) Self-awareness-based resource allocation strategy for containment of epidemic spreading. Complexity: 3256415. Available from https://www.hindawi.com/journals/complexity/2020/3256415/
8. Chen Y (2021) Analysis of occupation intergenerational mobility mechanism: based on the perspective of floating population and registered residence system. IOP Conf Ser Earth Environ Sci 692(4):42–80
9. Cheshmehzangi A (2020a) Reflection on disruptions: managing the city in need, saving the city in need. In: The city in need. Springer, Singapore, pp 137–283
10. Cheshmehzangi A (2020b) Recommendations for 'the city in need'. In: The city in need. Springer, Singapore, pp 285–304
11. Cheshmehzangi A (2020) COVID-19 and household energy implications: what are the main impacts on energy use? Heliyon 6(10):e05202
12. Cheshmehzangi A, Sedrez M, Ren J, Kong D, Shen Y, Bao S, Xu J, Su Z, Dawodu A (2021) The effect of mobility on the spread of COVID-19 in light of regional differences in the European Union. Sustainability 13(10):5395
13. Cheshmehzangi A, Li Y, Li H, Zhang S, Huang X, Chen X, Su Z, Sedrez M, Dawodu A (2022) A hierarchical study for urban statistical indicators on the prevalence of COVID-19 in Chinese city clusters based on multiple linear regression (MLR) and polynomial best subset regression (PBSR) analysis. Sci Rep 12(1):1–16
14. Esses VM (2018) Immigration, migration, and culture. In: Oxford research encyclopedia of psychology. Available fromhttps://doi.org/10.1093/acrefore/9780190236557.013.287
15. Galeazzi A, Cinelli M, Bonaccorsi G, Pierri F, Schmidt AL, Scala A, Quattrociocchi W (2020) Human mobility in response to COVID-19 in France, Italy and UK. Sci Rep 11(1):1–10
16. Gauvin L, Tizzoni M, Piaggesi S, Young A, Adler N, Verhulst S, Ferres L, Cattuto C (2020) Gender gaps in urban mobility. Human Soc Sci Commun 7(1):1–13
17. Geldsetzer P (2020) Use of rapid online surveys to assess people's perceptions during infectious disease outbreaks: a cross-sectional survey on COVID-19. J Med Internet Res 22(4):e18790
18. Hisi AN, Macau EE, Tizei LH (2019) The role of mobility in epidemic dynamics. Physica A 526:120663
19. Jhaveri R (2020) Echoes of 2009 H1N1 influenza pandemic in the COVID pandemic. Clin Ther 42(5):736–740

20. Kraemer MU, Yang CH, Gutierrez B, Wu CH, Klein B, Pigott DM, Scarpino SV (2020) The effect of human mobility and control measures on the COVID-19 epidemic in China. Science 368(6490):493–497
21. Li H, Cheshmehzangi A, Zhang Z, Su Z, Pourroostaei Ardakani S, Sedrez M, Dawodu A (2022) The correlation analysis between air quality and construction sites: evaluation in the urban environment during the COVID-19 pandemic. Sustainability 14(12):7075
22. Linka K, Goriely A, Kuhl E (2021) Global and local mobility as a barometer for COVID-19 dynamics. Biomech Model Mechanobiol 20(2):651–669
23. Liu Y, Mao B, Liang S, Yang JW, Lu HW, Chai YH, Wang L, Zhang L, Li QH, Zhao L, He Y (2020) Association between age and clinical characteristics and outcomes of COVID-19. Euro Respiratory J 55(5)
24. Pirouz B, Shaffiee Haghshenas S, Shaffiee Haghshenas S, Piro P (2020) Investigating a serious challenge in the sustainable development process: analysis of confirmed cases of COVID-19 (new type of coronavirus) through a binary classification using artificial intelligence and regression analysis. Sustainability 12(6):2427.
25. Pourroostaei Ardakani S, Xia T, Cheshmehzangi A, Zhang Z (2022) An urban-level prediction of lockdown measures impact on the prevalence of the COVID-19 pandemic. Genus 78(1):1–17
26. Pullano G, Valdano E, Scarpa N, Rubrichi S, Colizza V (2020a) Population mobility reductions during COVID-19 epidemic in France under lockdown. MedRxiv. Available from https://www.epicx-lab.com/uploads/9/6/9/4/9694133/inserm-covid-19_report_mobility_fr_lockdown-20200511.pdf
27. Pullano G, Valdano E, Scarpa N, Rubrichi S, Colizza V (2020) Evaluating the effect of demographic factors, socioeconomic factors, and risk aversion on mobility during the COVID-19 epidemic in France under lockdown: a population-based study. Lancet Digital Health 2(12):e638–e649
28. Puello LLP, Chowdhury S, Geurs K (2019) Using panel data for modelling duration dynamics of outdoor leisure activities. J Choice Modell 31:141–155
29. Rubinson L, O'Toole T (2005) Critical care during epidemics. Crit Care 9(4):1–3
30. Schwanen T, Dieleman FM, Dijst M (2001) Travel behaviour in Dutch monocentric and policentric urban systems. J Transp Geogr 9(3):173–186
31. STHDA (n.d.) Correlation matrix: a quick start guide to analyze, format and visualize a correlation matrix using R software—easy guides—Wiki. Available from http://www.sthda.com/english/wiki/correlation-matrix-a-quick-start-guide-to-analyze-format-and-visualize-a-correlation-matrix-using-r-software#what-is-correlation-matrix
32. Thorhauge M, Kassahun HT, Cherchi E, Haustein S (2020) Mobility needs, activity patterns and activity flexibility: how subjective and objective constraints influence mode choice. Transp Res Part A Policy Practice 139:255–272
33. TomTom Traffic Index (2020) As our world changes, traffic tells the story. Retrieved 7 May 2021. Available from https://www.tomtom.com/en_gb/traffic-index/
34. World Health Organization (WHO) (2020) Coronavirus disease 2019 (COVID-19) situation report—101. Availble from https://www.who.int/emergencies/diseases/novel-coronavirus-2019

Chapter 6
Adaptive Feature Selection for Google App Rating in Smart Urban Management: A Big Data Analysis Approach

6.1 Introduction

Big data analysis is prevalent in many different areas, including city transportation management, education, and biomedical sciences. In city management, big data use and analysis applications have grown popular, mostly aligned with smart city directions, smart urban systems, etc. With a large number of features inside the database (i.e., when the dataset size is too large), the whole analysis process could be time-consuming. Thus, it is essential to apply certain feature selection techniques to increase the machine learning speed in the data analysis stage. In addition, the machine learning model could be better trained by using a suitable feature selection method to select the most critical variables and eliminate redundant or irrelevant features. Most feature selection techniques focus on a similarity matrix that assigns a fixed value to pairs of objects. However, if the dataset is large and includes noise or missing samples, the possibility of having incorrect results would be higher. Thus, data analysis requires careful feature selection methods, which are highlighted in the discussions of this chapter. In this regard, a comparison analysis between adaptive feature selection methods and traditional feature selection techniques is provided.

This chapter introduces a study that compares the effectiveness of adaptive feature selection methods (e.g., Random Forest or RF) and traditional feature selection methods (e.g., Linear Discriminant Analysis or LDA, and Principal Component Analysis or PCA). In urban systems, such methods are commonly used. In this study, these three methods are applied separately during the feature selection phase, which helps us develop a comparative analysis as well. In doing so, we provide an opportunity to contrast their influence on predicting ratings in the Google Play Store dataset. The prediction accuracy for each category in the rating is then analysed to find out RF application in selecting features and whether this can outperform other methods such as PCA or LDA. This approach's broad applications could help optimize smart systems, particularly in cities.

The rest of this chapter is organized in the following. Section 6.2 of the chapter provides a brief survey of related work, focusing mainly on available and common

S. Pourroostaei Ardakani and A. Cheshmehzangi, *Big Data Analytics for Smart Urban Systems*, Urban Sustainability, https://doi.org/10.1007/978-981-99-5543-5_6

methods. We present the data pre-processing techniques in Sect. 6.3. In this section, we explore three key areas of data cleaning, data transformation, and data reduction, particularly for big data analysis at larger scales. There are broad applications for these three key areas, which are discussed briefly. The applications are visible for smart systems, particularly when dealing with larger areas of cities and regions. In Sect. 6.4 of this chapter, we delve into methods by exploring one adaptive feature selection technique and two traditional selection methods. These methods are used in the experiment conducted as part of the study. In Sect. 6.5 of the chapter, the experimental results are presented to verify the effectiveness of Random Forest (RF), which then lead to our brief conclusions and suggestions for future research.

6.2 Literature Review

6.2.1 Traditional Dimension Reduction Techniques

Since we need to use traditional dimension reduction algorithms for comparison, we also provide a review of a few literature studies about some of the relevant widely-used dimension reduction methods. In general, we can verify that there are two main methods of dimension reduction, i.e., 'Feature Selection (FS)' and 'Feature Extraction (FE'). Feature selection means no knowledge is required to select features from the original dataset. During feature selection, information may be lost due to the need to exclude certain features in the feature subset selection process. In feature extraction, the dimension can be reduced without losing much information [30] about the initial features.

Some widely used feature selection algorithms include 'Sequential Selection Algorithms (SSA)' and 'Heuristic Search Algorithms (HAS)'; both are wrapper approaches, which means they both apply the interpreter as a black box and use objective functions to maximize evaluations of current features [30]. In addition, there are also some other approaches, such as heuristic approaches, including Particle Swarm Optimization (PSO), Ant Colony Optimization (ACO), Genetic Algorithm (GAS), and others [33]. Feature extraction techniques could be divided into linear and nonlinear [5]. For linear methods, two of the most widely used techniques are 'Principle Component Analysis (PCA)' and 'Linear Discriminant Analysis (LDA)' [33]. Additionally, we provide a brief review of Kernel PCA, Multidimensional scaling (MDS), and Isometric Feature Mapping for nonlinear techniques. These methods all perform well with complex nonlinear data, which are highly applicable for larger-scale studies for cities and regions.

The choice between the Feature selection and Feature extraction algorithms is influenced by the particular dataset used in the project [30]. This could differ depending on other associated variables and the extent of the dataset. Similar to what [4] conducted in their LAP and ULAP research, we focus on feature extraction techniques to compare with our Random Forest method. In this respect, we picked

Principle Component Analysis (PCA) and Linear Discriminant Analysis (LDA) from feature extraction techniques to do some further review. These are summarized below.

(1) ***Principal Component Analysis***

Big data analysis is increasingly common, and such large datasets could sometimes be difficult to understand. Principle component analysis, commonly referred to as PCA is one of the most important and oldest techniques for feature selection [20]. This technique helps remove irrelevant features in a high-dimensional dataset without influencing the performance of data analysis. PCA is slightly different from some feature selection methods. It transforms the original feature set into a new feature set with a smaller number of features. In doing so, the transform result is not directly correlated to the feature component of the original dataset. Its main component is easy–finding new variables (i.e., principal components) based on the provided dataset to solve the eigenvalue and eigenvector problems, reducing the dimensionality of a dataset (i.e., usually the ones that have the highest eigenvalues), and at the same time, maintaining as much information as possible. PCA was first introduced by Pearson [23] and then by Hotelling [10], and a considerable number of scholarly works have been written on this subject [5, 13, 14, 15, 32]. PCA is a powerful feature extraction method, catching most of the information from data [35].

However, compared to adaptive feature selection methods such as RF or SVM, PCA has various limitations. These limitations are mainly based on the following three areas. First, PCA assumes there are correlations between features. If such correlations do not exist, then PCA cannot discover principal components. Second, PCA can be different when the scale of the data changes. That is to say, if the features are not standardized, PCA will tend to select features with a larger range regardless of the actual maximum variance. And third, PCA is not suitable for dealing with features with non-linear relationships as it assumes there is only a linear relationship between each other [12].

(2) ***Linear Discriminant Analysis***

Another classical and commonly-used method for data classification and dimensionality reduction is 'Linear Discriminant Analysis (LDA)'. It is a supervised dimension reduction technique, and the original technique was developed in the year 1936 by Ronald A. Fisher. Therefore, it is also named 'Fisher Linear Discriminant (FLD)'. Compared to PCA, which does more feature classification, LDA does more data classification. This technique tries to maximize the ratio of the difference between classes while minimizing the variance within the class, so it is always used in the field of pattern recognition [29]. Despite everything, the goal of LDA is to project the features in higher dimensional space onto a lower-dimensional space and reduce resources and dimensional costs. Take a two-class problem in a 2D graph as an example, shown in Fig. 6.1, we can verify that Linear Discriminant Analysis tries to find a line in which groups separate all the projections on the line. The distance between the centers of the two groups should be maximized, and the distance within one group of data should be kept as small as possible.

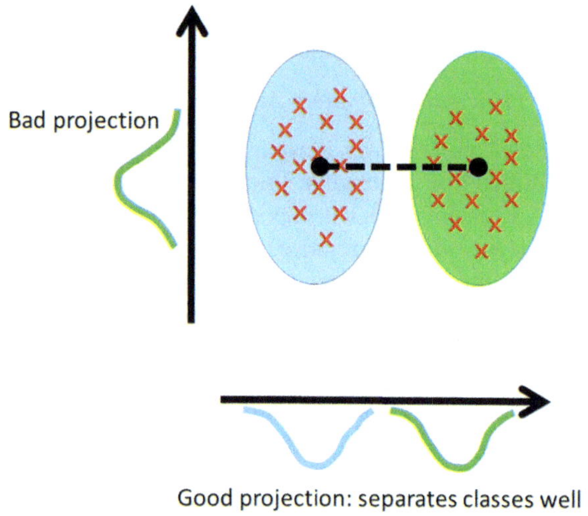

Fig. 6.1 Two-class problem in a 2D graph using linear discriminant analysis or LDA

When using LDA, it is assumed that all features in the dataset are Gaussian distributions with the same variance, and they are all randomly sampled. This means each feature is shaped like a bell-shaped curve [11]. In this regard, LDA is also one of the most common multivariate statistical data analysis tools. It generally performs much better than PCA for a large sum of data. This technique is applied in many fields, like bankruptcy prediction, face recognition, marketing, biomedical studies, earth science, positioning, and product management. Thus, its broad application is suitable for studying urban systems and optimizing the city/urban operations.

6.2.2 *Random Forest*

Random Forest (RF) is an integral classification algorithm based on the decision tree model. The decision tree is one of the least complex algorithms in classifying problems. Instead of relying on general similarities (e.g., 'Euclidean distance' and 'Hamming distance') to make decisions, the decision tree is able to categorize the data by thresholds in each non-leaf node, and each of them contains judgments based on one attribute [31]. To minimize the generated decision tree, entropy, which measures the impurity of input samples, is introduced to decide which attribute should be applied to a certain node. The feature with high entropy reduction indicates that it is effective in reducing the impurity of input data, which is the general idea of feature selection in RF [25]. RF algorithm can be seen as an ensemble of Decision Trees that focus on different attributes or features respectively. This approach has a pervasive application because it has good stability and generalization [36]. Therefore, it is effective for larger-scale studies, such as smart urban systems [6]. Moreover, since

RF does not rely on traditional similarities, it is not sensitive to scale operations such as normalization and standardization.

Before exploring RF further, it is recommended to do research on 'Bagging' [31]. With a dataset containing *m* sample importing into the Bagging algorithm, it is noted that a sample is randomly selected to be put into the sampling set. Then the data is returned to the initial dataset, which makes the next selection of this sample possible. After m rounds of random sampling, *m* sampling set with sample data is obtained. By analyzing the data, some of the samples occurred several times in the initial training set, while others did not [19]. Those that did not appear could be applied for the verification and later evaluation.

RF can be applied in feature selection tasks, since the importance of each feature can be calculated [16]. In this study, we focus on obtaining impurity-based importance. In RF, the focus could be on impurity-based importance, either of Gini or Entropy, which can be utilized to measure the impurity [3, 36]. Based on the analysis, there are two random aspects: (1) Bagging sampling and (2) the random-selected features for each node. In the calculating process, the impurity decrease is calculated and weighted by the number of data objects on that node and summed for the same appearing features in one internal decision tree [17]. Then, the average values of each feature, representing the importance, are attained among all decision trees. Finally, they are normalized to make the sum of feature importance equal to 1.

Admittedly, several drawbacks appear in RF [17]. First, the impurity-based importance seems to be not reliable if existing features that include a high number of categories [28] lead to the inflation of the importance of some features. Second, since highly-correlated features will share the importance in the impurity-based method, their importance will be underestimated and hard to detect [7]. RF is built based on the Decision Tree to construct the Bagging integration and additionally adds the selection of random attributes to the training process of the decision tree [8]. In RF, for each base node in the decision tree, a subset of the tree that stores *k* attributes is randomly picked from the set of candidate attributes from this node primarily. Later, from this subset, an optimal attribute is selected for division [16]. The choice of the number of attributes *k* to be extracted is important and is generally recommended. Eventually, the final decision is made by adopting the decision of majorities based on voting among decision trees. The application of RF could effectively handle big data as it is highly parallelized [3]. Also, sampling the decision tree candidate division attributes allows for efficient training of the model despite the high dimensionality of the sample features. The importance of each feature to the prediction can be calculated, and even the loss of partial features is tolerated [22]. However, the features that take more divisions of values tend to have a more significant impact on the decisions of the node, thus on the effectiveness of the fitted model [19]. While applying RF to the adaptive feature selection, the parameter setting and the feature selection seemed to be the most significant.

6.2.3 *Data Pre-processing*

This section is divided into three parts, focused on 'Data Cleaning', 'Data Transformation', and 'Data Reduction' [9]. This section presents data pre-processing techniques and corresponding examples that are relevant to this research study and other similar studies of urban systems. We note that PySpark can handle large data sets faster than Pandas and has other advantages, such as error tolerance [27]. Thus, all the techniques used in this section are based on the PySpark and related libraries.

A. ***Data Cleaning***

(1) *Missing data handling:* Since we need to handle a huge amount of data, we disregarded all rows with missing data [9]. And for the predicted label 'rating', we disregarded all rows with rating '0'. This is part of the data selection process for data handling, in particular, in dealing with missing data or information.

(2) *Noise data handling:* We chose to disregard all rows with meaningless data that cannot be interpreted by machines [9]. These data include examples like 'Varies with device' in attribute 'Size' and 'Unrated' in attribute 'Content Rating'. Again, this is part of the data selection and screening process, commonly practiced in the data handling process.

B. ***Data Transformation***

(1) *Number Hierarchy Generation:* In order to simplify the model and improve the training efficiency, we decided to convert the label 'rating' from a lower level to a higher level in the hierarchy [9]. The hierarchy rules are shown as all 'rating' data, which then discard the data after the decimal point and treat the data before the decimal point as the final label.

(2) *Data type transformation:* The original types of data were all saved as 'String', which was not conducive to model training and testing. Hence, we converted the data types according to the corresponding data attributes. For example, for attribute 'Price', we transformed the data type from 'String' to 'Double'.

(3) *Data unit transformation:* The original data in attribute 'Size' was not stored in uniform units. Hence, we unified the units into 'MB' and made the necessary conversions.

(4) *Time data transformation:* The time data in the original data were stored in the format of month-day-year. The data was too scattered, and the training efficiency was not high. Hence, we retained the data of year, and dropped the data for month and day.

C. ***Data Reduction***

(1) *Attribute Subset Selection:* The highly relevant attributes should be used, and the rest can be discarded. We dropped the data that contained less information for our classification task. Including' App Name', 'App ID',

‘Currency’, ‘Minimum Android’, ‘Developer ID’, ‘Developer Website’, ‘Developer Email’, ‘Privacy Policy’, ‘Scraped Time’, and ‘Installs’.

(2) *Row Subset Selection:* There was a serious data imbalance problem that occurred in some attributes, which is a small part of the data. This was noted as a minor imbalance compared with the overall data. This was not conducive to efficient model training. Hence, we dropped the data that were under some certain defined threshold. For example, for the attribute “Rating Count”, we dropped the data smaller or equal to 10.

(3) *Label Subset Selection:* We noticed that in the original data, after preliminary data processing, the number of labels of one type was much more than that of others, which was easy to cause model over-fitting [2]. Therefore, we filtered the data according to the ratio of labels in the original data to ensure that the ratio of each label in the data was roughly the same.

6.3 Methodology

After feature selection, different training feature sets were input into Random Forest (RF) classier to generate machine learning models. During the training phase, parameters for the classifier were set to be consistent, as shown in the code file.

A. *Principal Component Analysis*

During the experiment, *PCA*, the inbuilt function from MLlib, is used to do feature extraction, which is to calculate the principal component using the correlation method [24].

B. *Linear Discriminant Analysis*

Since “spark” does not provide the function of Linear Discriminant Analysis (LDA), we turned to use “Sklearn” for the study.

C. *Random Forest*

The Random Forest (RF) algorithm was selected as the main approach to identify the importance of each feature for selection and calculating the overall accuracy in this study.

In this study, we train the model for the prediction of ratings in the Google Play Store. There are 13 features for each data item, while RF is able to use the features for selection and finally get a rating prediction. Randomness is applied while selecting the best point for splitting in the RF. Here, ***F*** stands for the number of features in the dataset [8]. Algorithm 6.1 describes how the RF algorithm works with ***N*** as the number of training sets, ***S*** for the dataset, and ***M*** for the number of features in the dataset.

Algorithm 6.1 Random Forest Algorithm

Input N, S
Create an empty vector $R\vec{F}$
for $i = 1 \rightarrow N$ **do**

Create an empty tree T_i
repeat

Sample S out of all features F using Bootstrap sampling
Create a vector of the sample S features $\vec{F}s$
Find Best Split Feature $B(\vec{F}s)$
create A New Node using $B(\vec{F}s)$ in T_i

until No more Instances To Split On
Add T_i to the $R\vec{F}$

end for

Even if the RF can reach the goal, the overall performance could be improved by applying the importance of features in the dataset [26]. In the decision tree algorithm, the single trees are highly interpretable. However, this feature becomes disoriented in the RF, leading to difficulty in evaluating the contribution of different features to the model. Therefore, the performance would decrease with the disturbance from the useless features [8]. To fix this issue, the importance of features should be measured. On the other hand, by calculating the importance of features in RF, the overall performance and the accuracy of the model will be improved. Therefore, the importance of each feature could be computed [1]. Consequently, in Algorithm IV-C, the importance of each feature is calculated.

Algorithm 6.2 Feature Importance in Random Forest Algorithm

for features x_j, $j = 1$ to p **do**

for tree base learners $\hat{b}^{[m]}(x)$, $m = 1$ to M **do**

Find all nodes N in $\hat{b}^{[m]}(x)$ that uses x_j.
Compute improvement in splitting criterion achieved by them.
Add up these improvements.

end for

Add up improvements over all tress to get feature importance of x_j.

end for

Following the abovementioned algorithm, we design our own RF algorithm, which is then applied to the model. Initially, the feature 'Rating' was selected for the data model. Based on this model, the rating label was selected to be predicted. At this stage, a predicted label was gained with the counts of training instance labels from the tree node for prediction and raw prediction vectors normalized to a multinomial distribution. Then the *feature Importances* function was called to compute the mean and standard deviation of accumulation of the impurity decrease within each tree. The data was stored as a vector. For further calculation, it was converted to a list in the type of float. Before sorting the importance under a descending order and conducting impurity reduction, the list experienced another type of transformation from list to dictionary. Therefore, with the sorted list, the top eight of the most influential features were selected as the new features for prediction. Before creating the test and training dataset, the dataset merged multiple columns into a vector column. The RF classifier algorithm was called here in training the data on the training dataset, while evaluation was conducted on another test dataset. The data was prepared from two feature transformers, which allow indexing categories for the label and categorical features. The model output was printed here with features, ratings, raw prediction, probability, and prediction. The table revealed an overview of prediction and the relationship between each attribute. In doing so, we could calculate accuracy on evaluating the final performance of this RF algorithm.

6.4 Results and Discussions

6.4.1 Overall Comparison

Initially, we adopt the unbalanced dataset to get the prediction results, which would then be used for later comparison analysis. The accuracy seemed to be reliable at 63% in the five-classification problem. Nevertheless, all predicted values are 4. This phenomenon is caused by the serious data imbalance in predicting labels in the dataset. We found that the total number of samples in category four exceeded half a million, while the total number of samples in category one was only 7822. Therefore, we produced a new dataset sampling from the old one to make the proportions of each label nearly the same. The accuracy of random forest dropped back to approximately 40%. Since only a simple standardization was applied to all attributes, including both categorical and numeric features, we altered several scaling methods on the dataset. These scaling methods include 'Min–Max Normalization' and 'Z-Score Standardization' on numeric features. In addition, we utilized the 'one-hot encoding' on categorical features. After several rounds of intensive practicing, including adjusting parameters like the number of internal decision trees, the results from all models did not change and remained in the range of 1%. The imbalance of other features might cause low accuracy except for the label. This is due to the fact that in the previous balancing process, only the label feature was balanced. Thus, the imbalance of other features might further deteriorate in the sampling. The final results are shown in Table 6.1.

The other aspect of the study is to consider accuracy, which is very important for big data analysis at such a scale. Because we used the same RF method to

Table 6.1 Accuracy for each rating category

Name	1	2	3	4	5	Total
PCA	17.87	58.64	21.29	7.81	83.47	37.18
LDA	28.24	25.67	26.89	26.86	37.67	29.13
RF	56.27	4.44	31.41	34.52	78.60	41.41

train and classify the datasets, RF's feature selection obviously performs better than PCA and LDA in feature selections. Therefore, it is worth noticing that in Table 6.1, the accuracy values of Category 2 in RF and Category 4 in PCA are extremely low, and Category 5 has the highest accuracy values. The dataset has been balanced before to ensure all categories in rating have similar proportions. One possible reason for this phenomenon is that there are different categories between the train and test datasets. Figure 6.2 displays the importance of each feature under different situations. The features are 'Ad Supported', 'Category', 'Content Rating', 'Editors Choice', 'Free', 'In-App Purchases', 'Last Updated', 'Maximum Installs', 'Minimum Installs', 'Price', 'Rating Count', 'Released', and 'Size' from left to right.

6.4.2 *Discussion on Random Forest*

For feature selection in RF, as highlighted in Fig. 6.2, only three features' importance is over 0.01. This finding explains the reason behind the occurrence of low accuracy. The three most important features are 'Maximum Installs', 'Ad Supported', and'Rating Count'. Surprisingly, the importance of 'Category' is in the bottom five since, intuitively, 'Category' should be one of the most important features used to distinguish the rating. Thus, it is important to verify the influence of highly correlated features on the calculation of RF feature importance (i.e., in the pre-processing stage). This means that at this stage, two pairs of features with high correlations are preserved: (1) 'Free' and 'Price', and (2) 'Minimum Installs' and 'Maximum Installs'. Based on the exploration in the literature review, the highly correlated features will share the importance, making them underestimated. In the original dataset produced in the pre-processing stage, the importance of the 'Price' was 0.0024. After dropping the 'Free' column, it raised to 0.0032. The importance of 'Maximum Installs' was 0.30. However, deleting the'Minimum Installs' column raised it to 0.42.

According to Fig. 6.2, selecting only the three most critical features is enough since their importance values far exceed the importance of other features. We applied tenfold cross-validation to the RF algorithm to verify from this perspective. We applied a different number of selected features for each fold and attained accuracy. The results are displayed in Fig. 6.3. According to the results, after choosing more than the three most essential features, the line chart is flattened out and fluctuates

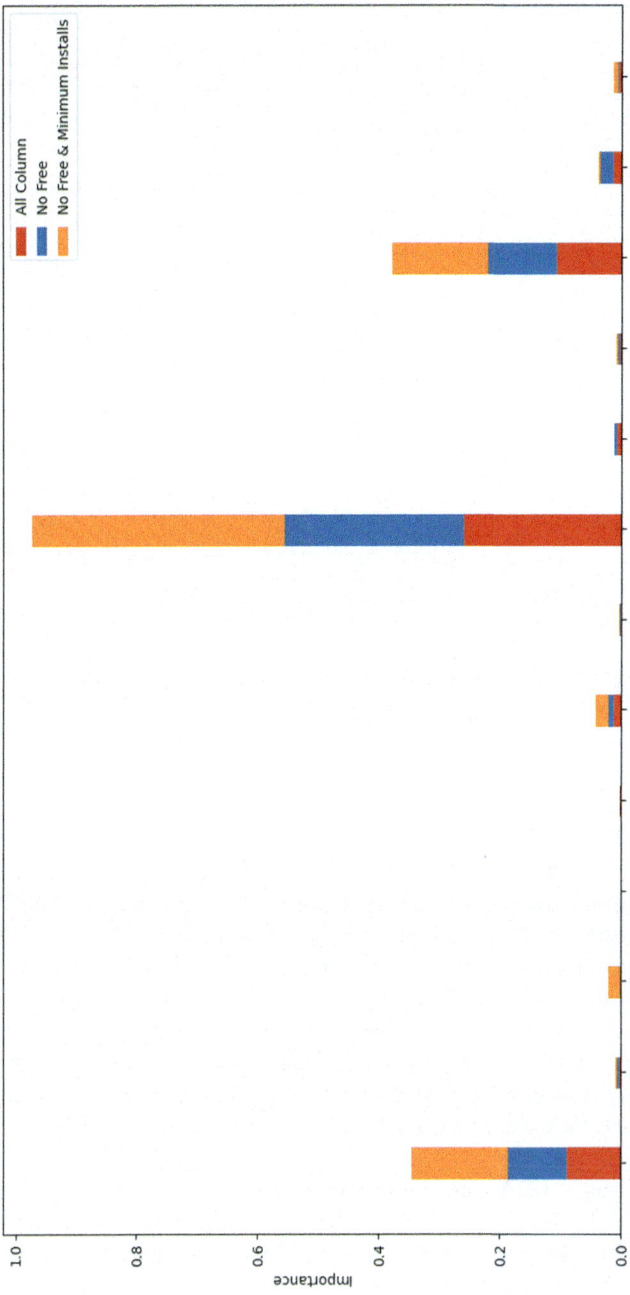

Fig. 6.2 RF Feature importance in different situations

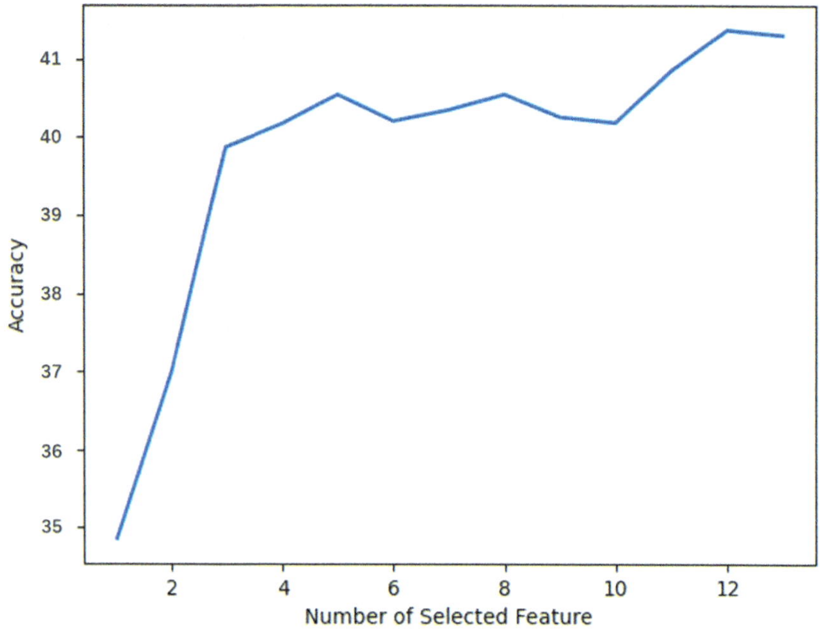

Fig. 6.3 RF Accuracy in different selected features

around 0.40 and 0.42. Thus, 'Maximum Installs', 'Ad Supported', and'Rating Count' may be the main components to determine the rating.

6.4.3 Discussion on Linear Discriminant Analysis

As demonstrated in Table 6.1, in LDA, the accuracy for predicting labels "1" and label "5" was better than the accuracy compared to the other three. It is because LDA tries to maximize the distance between each label in the projection, and the between-class scatter matrix completely relies on the calculation of the total data mean. Actually, it is very easy to happen with LDA that edge classes, which are classes with larger deviations, can be in dominant places compared to other classes [35]. The features of some edge-label data may be predicted to be less than or greater than the edge, and they were finally labeled as in the edge classes after prediction [18].

The final output of LDA was worse than we had expected. The reason for this result is based on the fact that LDA assumes that the input data matches a Gaussian distribution, which means each feature is shaped like a bell-shaped curve. However, our data had no obvious pattern of Gaussian distribution. In addition, although LDA is generally thought to be better for big data problems than PCA [21], it is seen that, in this case, it performed worse than PCA. This result is because LDA relies more on the mean value, not the variance.

On the other hand, it can be seen clearly that the variance of the accuracy of each label using LDA was less than that of using RF and PCA. This means the fluctuation between accuracy for each label was relatively small. This finding can be a hint for improving the performance of the other methodology in future research studies of big data analysis.

A. *Discussion on Principal Component Analysis*

As we mentioned in the literature review, although PCA is a powerful tool for feature extraction, it is possible that information could be lost during this process. In its standard equation, PCA is a linear equation, where X and Y are the input and output vectors, and W is the matrix that transforms original features:

$$Y = W^T X \tag{6.1}$$

In this case, PCA should be invertible. However, the output vector is truncated to reduce the original feature dimensions; thus, information is lost during this process.

B. *Further Improvement*

Even though the experiment showed that the adaptive feature selection method (i.e., RF) could achieve higher performance in data analysis compared to traditional feature selection techniques (e.g., PCA, LDA), we can verify that some future work can be done. Due to time and resource limits, we only use one dataset to demonstrate that RF can help achieve higher accuracy in big data analysis compared to PCA and LDA. Thus, further research is needed to establish stronger proof on this topic. In addition, further studies should take into account more machine learning techniques apart from random forest classifier as we used in our experiments so that the robustness of our proposal could be further proved.

6.5 Conclusions

In this chapter, we have compared the adaptive feature selection technique (Random Forest (RF) in this case) with two well-known traditional feature selection techniques (i.e., PCA and LDA). Random Forest builds on the Decision Tree to evaluate the importance of each feature for selection and calculate the overall accuracy. We performed RF on the "Google-playstore"dataset to test the feasibility of using local adaptive feature techniques on big data. The label we chose to predict was "Rating", and we had labeled it so the whole machine learning procedure was decided to be a classification.

In the first stage, we used data preprocessing with spark to better use the dataset. In data preprocessing, the data columns that we agreed to be unnecessarily related to "Rating" were dropped. Then, each column was cleaned by dropping the rows containing null or empty values, and the data type of each column was changed to numbers or Boolean so they could be used in the machine-learning procedure. We also noticed that the data was unbalanced. The majority of the data were in category 4, which would significantly affect the final output of training. That was why we also added one step to ensure each category's number for "Rating" was the same. Eventually, the data was split into 80% training data and 20% testing data. After that, the cleaned data was trained with different feature selection techniques, and the goal was to get the best accuracy for the model.

In RF, the importance of the features was calculated with the *feature Importances* function and input variables of an Integer, indicating the number of chosen features to be further used and the original dataset with all features. Another RF training would be called with the chosen labels to train a model for the final output, and the model's accuracy would be calculated with the test dataset. The PCA and LDA were also used independently to extract the dataset's key features. They also used RF to train the data to ensure the only difference in the experiment was the feature selection method. The result turned out that the RF performed much better than the other two, which demonstrated that RF was indeed a better feature selection technique in this case.

The findings from this study conclude that the adaptive feature selection gives optimized results compared to traditional ones. However, the result was still worse than we initially expected. It is suggested that future research could explore ways to improve the result of RF by changing the algorithm for calculating the importance of combining RF with other feature selection techniques. Such applications would be highly important for future big data analysis for urban systems and regional-level studies.

Acknowledgments The research work in this chapter was supported by our research team members: Meitong Wang, Yuan Dai, Deze Zhu, Peiwang Liu, and Zhanpeng Wang.

References

1. Abdoh SF, Abo Rizka M, Maghraby FA (2018) Cervical cancer diagnosis using random forest classifier with SMOTE and feature reduction techniques. IEEE Access 6:59475–59485. https://doi.org/10.1109/ACCESS.2018.2874063
2. Brownlee J (2022) 8 Tactics to combat imbalanced classes in your machine learning dataset. Mach Learn Mastery. [Online]. Available: https://machinelearningmastery.com/tactics-to-combat-imbalanced-classes-in-your-machine-learning-dataset/
3. Chen J, Li K, Tang Z, Bilal K, Yu S, Weng C, Li K (2017) A parallel random forest algorithm for big data in a spark cloud computing environment. IEEE Trans Parallel Distrib Syst 28(4):919–933. Available: https://doi.org/10.1109/tpds.2016.2603511

4. Chen X, Yuan G, Wang W, Nie F, Chang X, Huang J (2018) Local adaptive projection framework for feature selection of labeled and unlabeled data. IEEE Trans Neural Netw Learn Syst 29(12):6362–6373. Available from: https://doi.org/10.1109/tnnls.2018.2830186
5. Cheshmehzangi A, Li Y, Li H, Zhang S, Huang X, Chen X, Su Z, Sedrez M, Dawodu A (2021) A hierarchical study for urban statistical indicators on the prevalence of COVID-19 in Chinese city clusters based on multiple linear regression (MLR) and polynomial best subset regression (PBSR) analysis. Sci Rep 12, Article Number 1964. https://doi.org/10.1038/s41598-022-05859-8
6. Cheshmehzangi A, Pourroostaei Ardakani S (2021) Urban traffic optimization based on modeling analysis of sector-based time variable: the case of simulated Ningbo, China. Front Sustainab Cities 3, Article Number: 629940. https://doi.org/10.3389/frsc.2021.629940
7. Darst B, Malecki K, Engelman C (2018) Using recursive feature elimination in random forest to account for correlated variables in high dimensional data. BMC Genetics 19(S1)
8. Fawagreh K, Gaber M, Elyan E (2014) Random forests: from early developments to recent advancements. Syst Sci Control Eng 2(1):602–609. Available: https://doi.org/10.1080/21642583.2014.956265
9. GeeksforGeek (2021) Data preprocessing in data mining. Available at: https://www.geeksforgeeks.org/data-preprocessing-in-data-mining/
10. Hotelling H (1933) Analysis of a complex of statistical variables into principal components. J Edu Psycho 24(6):417–441. Available from: https://doi.org/10.1037/h0071325
11. Hussein Ali A, Faiz Hussain Z, Abd SN (2020) Big data classification efficiency based on linear discriminant analysis. Iraqi J Comput Sci Math 7–12. Available: https://doi.org/10.52866/ijcsm.2019.01.01.001
12. Kaboola.com (2022) A guide to principal component analysis (PCA) for machine learning. Online source, Available from: https://www.keboola.com/blog/pca-machine-learning
13. Karhunen J, Joutsensalo J (1995) Generalizations of principal component analysis, optimization problems, and neural networks. Neural Netw 8(4):549–562. Available from: https://doi.org/10.1016/0893-6080(94)00098-7
14. Kocherlakota S, Kocherlakota K, Flury B (1989) Common principal components and related multivariate models. Biometrics 45(4):1338. Available from: https://doi.org/10.2307/2531792
15. Leigh S, Jackson J (1993) A user's guide to principal components. Technometrics 35(1):84. Available: https://doi.org/10.2307/1269292
16. Lin W, Wu Z, Lin L, Wen A, Li J (2017) An ensemble random forest algorithm for insurance big data analysis. IEEE Access 5:16568–16575. https://doi.org/10.1109/ACCESS.2017.2738069
17. Liu Y (2014) Random forest algorithm in big data environment. Comput Model New Technol 18(12A):147–151
18. Lugosi G, Mendelson S (2021) Robust multivariate mean estimation: the optimality of trimmed mean. Annals Stat 49(1). Available from: https://doi.org/10.1214/20-aos1961
19. Lulli A, Oneto L, Anguita D (2019) Mining big data with random forests. Cognit Comput 11(2):294–316. Available: https://doi.org/10.1007/s12559-018-9615-4
20. Maćkiewicz A, Ratajczak W (1993) Principal components analysis (PCA). Comput Geosci 19(3):303–342
21. Martinez, Kak A (2001) PCA versus LDA. IEEE Trans Pattern Anal Mach Intell 23(2):228–233. Available from: https://doi.org/10.1109/34.908974
22. Melo CFOR, Navarro LC, de Olivera DN, Guerreiro TM, de Oliveira Lima E, Delafiori J, Dabaja MZ, et al (2018) A machine learning application based in random forest for integrating mass spectrometry-based metabolomic data: a simple screening method for patients with Zika Virus. Front Bioeng Biotechnol 6. Available from: https://doi.org/10.3389/fbioe.2018.00031
23. Pearson K (1901) On lines and planes of closest fit to systems of points in space. London, Edinburgh, Dublin Philosoph Mag J Sci 2(11):559–572. Available from: https://doi.org/10.1080/14786440109462720
24. Pham H (2007) Springer handbook of engineering statistics. Springer, New York
25. Reddy GT, Kumar Reddy MP, Lakshmanna K, Kaluri R, Rajput DS, Srivastava G, Baker T (2020) Analysis of dimensionality reduction techniques on big data. IEEE Access 8:54776–54788. https://doi.org/10.1109/ACCESS.2020.2980942

26. Rogers J, Gunn S (2006) Identifying feature relevance using a random forest, subspace, latent structure and feature selection, pp 173–184. Available from: https://doi.org/10.1007/11752790-12
27. Sparkbyexamples (2022) Pandas vs PySpark data frame with examples. Available from: https://sparkbyexamples.com/pyspark/pandas-vs-pyspark-dataframe-with-examples/
28. Strobl C, Boulesteix A, Zeileis A, Hothorn T (2007) Bias in random forest variable importance measures: illustrations, sources and a solution. BMC Bioinf 8(1)
29. Tharwat A, Gaber T, Ibrahim A, Hassanien A (2017) Linear discriminant analysis: a detailed tutorial. AI Commun 30(2):169–190. Available from: https://doi.org/10.3233/aic-170729
30. Velliangiri S, Alagumuthukrishnan S, Thankumar Joseph S (2019) A review of dimensionality reduction techniques for efficient computation. Procedia Comput Sci 165:104–111. Available from: https://doi.org/10.1016/j.procs.2020.01.079
31. Vens C (2013) Random forest, encyclopaedia of systems biology, pp 1812–1813. Available: https://doi.org/10.1007/978-1-4419-9863-7-612
32. Vidal R (2018) Generalized principal component analysis. Springer, Germany
33. Xu X, Liang T, Zhu J, Zheng D, Sun T (2019) Review of classical dimensionality reduction and sample selection methods for large-scale data processing. Neurocomputing 328:5–15. Available from: https://doi.org/10.1016/j.neucom.2018.02.100
34. Xu Y, Zhang D, Yang J (2010) A feature extraction method for use with bimodal biometrics. Pattern Recogn 43(3):1106–1115. Available from: https://doi.org/10.1016/j.patcog.2009.09.013
35. Yan C, et al (2021) Self-weighted robust LDA for multiclass classification with edge classes. ACM Trans Intell Syst Technol 12(1):1–19. Available: https://doi.org/10.1145/3418284
36. Zhou Q, Zhou H, Li T (2016) Cost-sensitive feature selection using random forest: Selecting low-cost subsets of informative features. Knowledge-Based Syst 95:1–11. Available: https://doi.org/10.1016/j.knosys.2015.11.010

Chapter 7
Improve the Daily Societal Operations Using Credit Fraud Detection: A Big Data Classification Solution

7.1 Introduction: An Overview of Recent and Ongoing Research on Credit Fraud Detection

With the rapid development of online shopping, online transactions have increased significantly. Unfortunately, the cases of online fraud are also improved. Based on Global Payments Report 2015, applying credit card is the most prevalent payment method worldwide compared with other payment methods, such as e-wallets and bank transfers [30]. Therefore, it is reasonable to infer that credit card fraud is the most popular item among other frauds since users' personal information could easily be revealed by hackers [10], which includes card numbers, CVV codes, passwords, etc. In general, credit card fraud could be categorized into three types, conventional fraud, such as stealing a physical card. Online frauds, for instance, faking the merchant sites by using the sensitive credit card information hacked from the users, and merchant-related frauds [32], colluding with the merchants to swindle customers.

Furthermore, credit card fraud negatively affects global finance, supported by observable data. In 2013, US retailers lost nearly 23 billion dollars due to credit card fraud, and the cost increased to around 32 billion dollars in the next year [3]. One of the significant causes of credit card fraud is the weak security of users' credit cards. In recent years, this issue has been trending alarmingly. According to the Nilson Report, the financial loss of credit card fraud will exceed 35 billion dollars approximately in 2020 [4]. Thus, it is necessary to develop sophisticated and frequently updated credit card fraud detection methods to reduce the tremendous loss.

Data mining and machine learning methods to detect fraud are the most popular among those methods. Data mining techniques could be applied to extract the most important information from a large-scale dataset, which involves statistics and mathematics to distinguish normal or abnormal transition information [32]. Nevertheless, data mining techniques could only discover valuable information. By compensating with machine learning techniques, the overall method could construct suitable models intelligently detect fraud.

S. Pourroostaei Ardakani and A. Cheshmehzangi, *Big Data Analytics for Smart Urban Systems*, Urban Sustainability, https://doi.org/10.1007/978-981-99-5543-5_7

In this chapter, we first apply data mining techniques to clean the raw data. Then, we perform feature engineering, such as Principle Component Analysis (PCA), to extract significant features. For experiment efficiency, we use Spark and its machine learning library, MLlib, to construct a pipeline. Several models are trained and tuned in order to classify the data with satisfactory accuracy. Then we investigate the efficiency and scalability of Spark while handling big data with more executors or cores.

In terms of our contributions to this project, different types of machine learning models were trained for this case, and the best model was selected as MLP with the highest accuracy. Moreover, the field of Spark's efficiency and scalability was investigated by experimenting with different sizes of datasets and a different number of executors and cores. We verified that the best performance could reach while the number of physical cores in the local machine equals the number of configured parallel degrees.

This chapter is structured as: Sect. 7.2 reviews the literature, and highlight the key research topics, Sect. 7.3 describes the research methodology, Sect. 7.4 presents and discusses the experimental results, and Sect. 7.5 concludes the discussions and highlights the future work.

7.2 Literature Review Related to Big Data and Credit Fraud Detection

The term big data is often used to describe massive datasets that are relatively hard to analyze with existing data processing tools [17]. Big data has great potential to increase efficiency and quality in industries such as healthcare, manufacturing, retail, and public administration, benefiting companies and customers [5]. However, processing big data has challenges and differences from traditional datasets in terms of variety, velocity, and volume [24]. This requires big data processing tools to be able to cope with big datasets from various sources in different structures and to perform all processes in a relatively short amount of time. Popular open source software for big data include MapReduce [7], Spark [34], and Hive [29].

Apache Spark is a popular big data analytics framework developed at UC Berkeley in 2009 and was made open source in 2010 [34]. Spark is integrated into Scala, Java, Python, SQL, and R. Spark has various upper-level libraries such as Spark SQL for structured data processing, Spark Streaming for stream processing, MLlib for machine learning, and GraphX for graph analysis available to perform tasks in different steps of big data analysis. These upper-level libraries are built upon Spark core, which implements the Resilient Distributed Dataset (RDD) abstraction for efficient processing of large-scale datasets. Spark monitors a graph of transformations to recover from failure and only computes RDDs when an action has been called, which means RDDs are fault tolerant and lazily evaluated [25]. In addition to RDDs, a DataFrame API [1] in Spark SQL has been developed to integrate relational data processing. A DataFrame represents a table in the relational database, and it can perform operations supported in relational databases, such as querying. Operations similar to those on RDDs and operations integrated with other libraries, including MLlib, are also implemented with DataFrames. The performance of Spark and MapReduce are compared on tasks including Word Count, Sort, K-means, and PageRank using datasets as large as 372 GB and 500 GB [26]. Their results showed that Spark is five times faster than MapReduce on K-means and PageRank and 2.5 times faster for WordCount, while MapReduce outperforms Spark on Sort. It was shown by Mostafaeipour et al. [19] that Spark could significantly outperform MapReduce in KNN classification on datasets of different sizes using the 8 GB Higgs dataset. Similar results were obtained by Gopalani and Arora [11] that Spark is twice as fast on K-means clustering compared to MapReduce.

MLlib is a Spark distributed machine learning library that aims to provide a solution for large-scale machine learning tasks by data and model parallelism, open sourced in 2013 [18]. MLlib implements common machine learning algorithms and supporting utilities with the Pipeline API and provides optimization for scalable machine learning. Additionally, it is also integrated with other Spark libraries such as Spark SQL. Spark and MLlib have been used in various studies. It was used to develop a Network Intrusion Detection System with the UNSW-NB15 public dataset [2]. A proposed sentiment analysis classification framework was implemented with Spark using two public Twitter datasets with over 1 million records [21]. The result shows that Spark is efficient and scalable for big data analysis. Panigrahi et al. [22] implemented a Hybrid Collaborative Filtering Recommender Engine with Spark using the Movielens dataset containing 20 million ratings. This model outperforms traditional collaborative filtering methods and improves scalability.

The survey done by Zojaji et al. [37] classified fraud detection techniques into two general categories: anomaly detection and misuse detection. Anomaly detection is based on an unsupervised methodology. In this method, the legitimate user behavioral model (e.g., user profile) for each account is required to be built, and then the fraudulent activities will be detected based on it. The implemented unsupervised models will detect abnormal data. External support from human experts is required to provide knowledge for the explanations of the abnormal data detected. The solution to the verified fraud pattern will be built into the detection model. Moreover, misuse detection deals with supervised classification tasks. The transactions are labeled as fraudulent or not fraudulent. There are numerous model creation methods for a typical two-class classification task, such as rule induction [9], decision trees [35], and neural networks [23]. The problem with this method is that it cannot identify innovative frauds. For the labeled data provided, this study adopts the misuse detection.

This project implements the big data technique with machine learning algorithms to implement credit fraud detection on the IEEE-CIS Fraud Detection public dataset from Kaggle [14]. This competition dataset was published two years ago. There have been a series of solutions proposed. According to the competition summary [31], the team winning the first prize implemented three main models: the Catboost,

LightGBM, and Xgboost. The Catboost, with a precision of 96.39% on the public testing data, is the best model for fraud detection. LGBM, with a precision of 96.17%, and XGB, with a precision of 96.02%, won the second position on the public and private test sets from Kaggle, respectively. Chen further implemented an improved Catboost [6] model and achieved an accuracy of 98.3% on this dataset. Lightgbm model raised by Ge et al. [10] achieved an accuracy of 98.2%. It beat the models of Xgboost, SVM, and logistic regression in its experiments. Zhang et al. [36] raised the Xgboost model, which achieved an accuracy of 98.1%, which is proved to perform better than the logistic regression with an accuracy of 93.1%, SVM with an accuracy of 95.8%, and the random forest model with an accuracy of 96.5%. Yu et al. [33] proposed a deep neural network model with an accuracy of 95.7%, and this NN model was verified to perform better than the SVM, followed by the logistic regression, and the Naïve Bayes was ranked as the last. The performances of the random forest and the NN model require further comparison.

In summary, the Catboost model was ranked as the best model, followed by the Lightgbm model. Xgboost is in third place. The random forest and NN models are better than SVM and the logistic regression models. Na¨ıve Bayes was ranked as the last.

7.3 Methodology

In this section, we highlight four main parts of (1) dataset introduction, (2) data preprocess and feature extraction, (3) model description, and (4) model implementation.

7.3.1 *Dataset Introduction*

The dataset used in this chapter is called IEEE fraud detection, extracted from Kaggle. The dataset has four major files: train_transaction.csv with 394 columns, train_identity.csv with 41 columns, test_transaction.csv with 393 columns, and test_identity.csv with 41 columns. The specific information of transaction files and identity files are shown in Tables 7.1 and 7.2. Key transaction IDs connect the transaction files and the identity files. But not all transaction IDs in transaction files can find matching transaction IDs in identity files. This problem is solved by using transaction IDs in the transaction files as the standard. After the combination, the training file has 433 features and 590,540 instances.

Table 7.1 Transaction information

Columns TransactionID	Description ID of transaction	Type ID
isFraud	Binary target	Categorical
TranscationDT	Transaction time	Time
TranscationAmt	Transaction payment amount	Numerical
ProductCD	Product code	Categorical
card1-card6	Payment card information	Categorical
addr1-addr2	Address	Categorical
dist1-dist2	Country distance	Numerical
P emaildomain	Purchaser email domain	Categorical
R emaildomain	Recipient email domain	Categorical
C1-C14	Counting (actual meaning is masked)	Numerical
D1-D15	Time delta (actual meaning is masked)	Numerical
M1-M9	Match (actual meaning is masked)	Categorical
V1-V339	Vesta features (actual meaning is masked)	Numerical

Table 7.2 Identification information

Columns TransactionID	Description ID of transaction	Type ID
Id_01-id_11	Identification data	Numerical
Id_12-id_38	Identification data	Categorical
DeviceType	Device type	Categorical
DeviceInfo	Device information	Categorical

7.3.2 Data Preprocess and Feature Extraction

The basic steps of data preprocessing are shown in Fig. 7.1. The first step is dropping features that have more than 80% of null values. The reason for choosing 80% is based on other people's experience processing this data set. Some people have tried to drop more than 87% of features that contain a certain amount of nulls [20]. However, some people have chosen to remove features that include more than 90% null, and the number of dropped features only accounts for about 3% of the total [8]. Thus, to ensure that a large amount of data is not lost and, at the same time, the number of meaningless features can be reduced as much as possible, 80% is our final choice. This dropping null step can remove 74 features. The second step is for C1-C14, D1-D15, and V1-V339 features. The purpose of this step is to reduce the number of special features and prevent the occurrence of overfitting. Take V1-V339 as an example. These 339 features have the same prefix, meaning they probably describe the same things. Based on this assumption, the number of such features can be greatly reduced by deleting highly correlated features and finding the representative features among them. To be more specific, calculate the correlation of each feature. If the correlation between two features is greater than 95%, delete one of the features [10]. This step can remove 98 features.

Through the first two steps, the remaining features are considered to affect the algorithm's performance to some extent. Before using these features, the null values among them must be dealt with first. Using −999 to make up those null values is feasible [6]. Although many features can be removed in the first two steps, there are still 261 features left, which will cost a lot of computation and space resources. We rely on PCA to solve this problem because PCA can reduce the dimensionality of the dataset and, at the same time, minimize information loss [13]. Table 7.1 and 7.2 refer to anonymous features including M1-M9, C1-C14, D1-D15, V1-V339, id_01-id_11, and id_2-id_38 which are unknown and meaningless. Although they are anonymous, features with the same prefix are related. Thus, they are further reduced using PCA. The principle behind this step is to put the features with the same prefix into the PCA and then use the specific number of features generated by the PCA to represent all the features of the prefix.

7.3.3 *Model Description*

(1) Support Vector Machine (SVM)

The model of performing classification using linear support vector machines (SVM) is also trained for credit card fraud detection. It is a binary classifier optimizing Hinge Loss by using the OWLQN optimizer. This model only supports L2 regularization currently. Linear separable datasets are required since the predicted results are not promising for the inseparable linear data. After randomly splitting the training and testing data to 7:3, the model tunning is performed by applying cross-validation with five folds. The highest accuracy for the 10 k dataset is 96.46%, with the maximum number of iterations 200 and the regularization parameter 1.

(2) Multilayer Perceptron (MLP)

A multilayer perceptron (MLP) is a feedforward artificial neural network with at least three layers: an input layer, a hidden layer, and an output layer. In this question, the number of input nodes and output nodes is fixed, 50 and 2, respectively. Thus, the number of hidden layers and the number of nodes in each layer determine this model's performance. Also, iteration times can affect the result to a certain extent. During the parameter tuning process, the maximum number of hidden layers tried is two because, for any desired shape, MLP with two hidden layers is sufficient to create classification areas [15]. The highest accuracy for the 10 k dataset is 96.98%, with the maximum number of iterations 500 and 1 hidden layer with seven nodes.

A. ***Logistic Regression***

Logistic regression is a generalized linear model used for classification problems by fitting a sigmoid function. Logistic regression is usually used

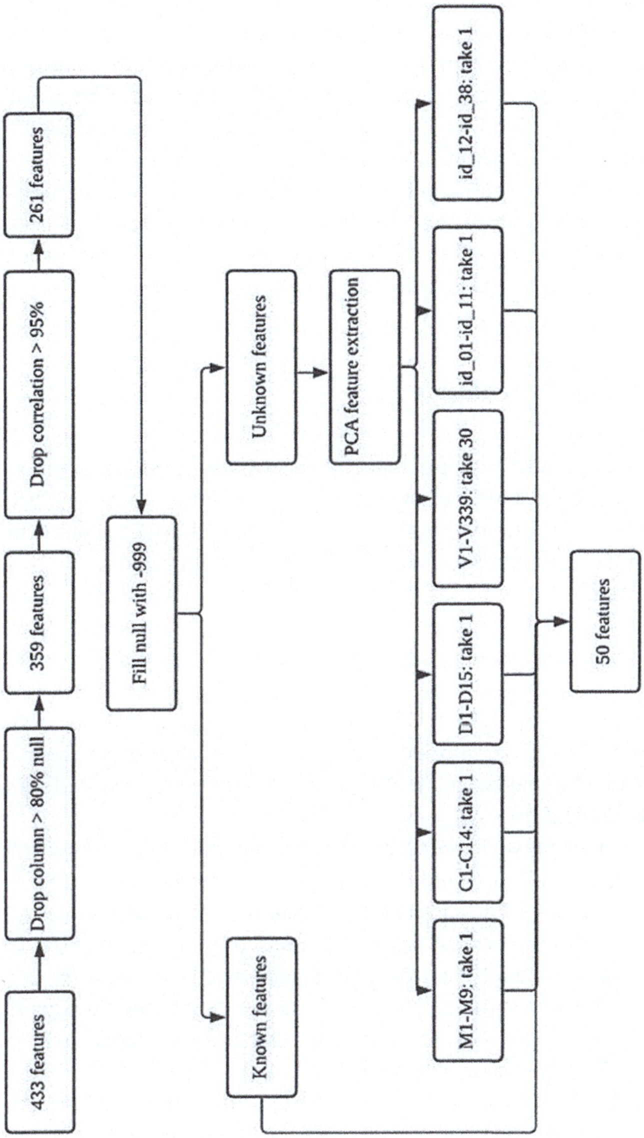

Fig. 7.1 Workflow of feature extraction

for binary classification by calculating the possibility of the two possible outcomes. This model is tuned in terms of several iterations, the regularization parameter, method, and penalty parameter. The best performance for the 10 k dataset is 96.21%, with L2 penalty and a regularization parameter of 0.01.

B. ***Random Forest***

Random forest is a type of ensemble learning method. Multiple decision trees form an ensemble in the random forest algorithm. The decision trees and their predictions are aggregated to produce the result. In random forest classification, the decision trees generate the classification result through a majority vote. The main hyper-parameters for the random forest model are the number of trees, the number of features sampled for each tree, and the depth of each tree. The highest accuracy for the 10 k dataset is 96.43%, with 30 trees and a maximum depth of 6 for each tree.

7.3.4 Model Implementation

The pre-processing, model training, and testing are combined into a pipeline. The four algorithms are tested and tuned using the five-fold cross-validation pipeline. The model with the best performance is selected as the model used for big data experiments. Different-sized datasets are sampled from the original dataset with respect to the ratio of the label classes to ensure the sampled datasets are valid. The sampled datasets contain 1, 10, 100, and 200 k records, respectively. This implementation is also tested regarding the number of executors and cores to find the best parallelization method using the 100 k dataset. The performances of the four implemented models are summarized in Table 7.3. The neural network achieved the best result on the testing dataset, and it was selected as the model for further experiments in III-E. Four executors in total and two cores for each executor are configured through *the spark-submit* command line.

The experiment results of driver program executing time are illustrated in Fig. 7.2. The original dataset is divided into different sizes of subsets whose unit is a kilo. The best tunning model, MLP, and its relative parameters are applied to this experiment. After enlarging the size of the test datasets from 1, 10, and 100 to 200 k, the execution time is gradually increased. It is acknowledged that if the correlation coefficient is significantly different from zero, we say their correlation coefficient is significant. Therefore, there is a significant linear relationship between the variables [12]. Analyzing the experiment results, it is rational to conclude that the relationship between distinguished data sizes and the execution time conforms to linear

Table 7.3 Model performance information

Model name	SVM	MLP	Logistic regression	Random forest
Accuracy (%)	96.46	96.98	96.21	96.43

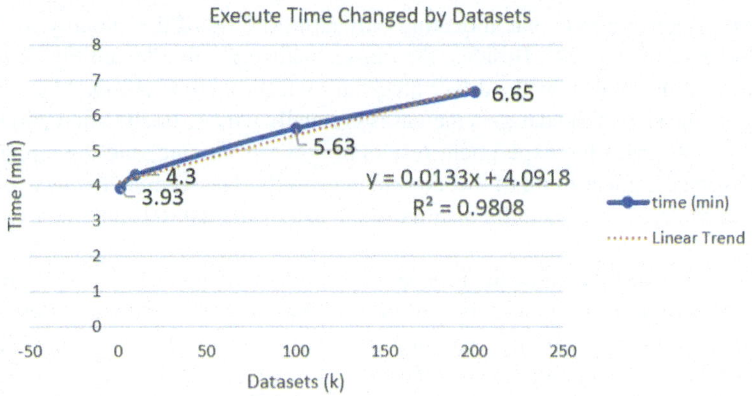

Fig. 7.2 This experiment measures how the driver program's execution time changed by different datasets' sizes. The size of the datasets increases gradually with 1, 10, 100, and 200 k. The blue line illustrates the experiment results in a unit of minute, and the orange line demonstrates a linear trend of the experiment results. The formula below expresses the linear relation between the distinguished size of the datasets and the execution time

regression since the input values obtain a p-value = 0.0096, which is less than the significant level of 0.05. Additionally, it is straightforward to calculate that R2 = 0.9808, demonstrating that two variables are positively correlated, conforming to our common sense that larger datasets require more processing time. Instead of corresponding to an exponential relationship, the processing time grows linearly with the increase of datasets. Thus, Spark obtains robust scalability to handle big data more efficiently with feasible time-consuming.

7.4 Results and Analysis

Apart from adopting and tunneling different machine learning models, we also attempt to verify Spark's efficiency in dealing with large datasets. As introduced before, Spark is suitable for big data since it is more efficient and scalable than its counterparts, such as MapReduce, concluded from the experiment implemented with Spark using two public Twitter datasets [21]. Since the provided datasets of fraud information are relatively large, with more than half a million instances, processing the data with an efficient tool, such as Spark, is significant. Accordingly, the datasets are also suitable for verifying the efficiency and scalability of Spark.

The experiments are performed on our local machine, which obtains Windows 10 Pro operating system and i7 CPU processors with 2.70 GHz, four physical processing cores, and eight logical processing cores. The most updated version, Spark-3.1.1, is used for the experiments under the environment of PYSPARK PYTHON. For the tests of Spark's scalability, we allocate 20 GB of memory for the driver program related to deployment. Then we distribute 1 GB to each executor. The Spark official documentation suggests that allocating only at most 75% of the memory for Spark should be sufficient [27]. Besides, the cluster manager will allocate Fig. 7.2. This experiment measures how the driver program's execution time changed by different datasets. The size of the datasets increases gradually with 1, 10, 100, and 200 k. The blue line illustrates the experiment results in a unit of minute, and the orange line demonstrates a linear trend of the experiment results. The formula below expresses the linear relation between the distinguished size of the datasets and the execution time.

Then we investigate how the number of executors affects the processing time of the driver program. By configuring a different number of executors and two cores for each, their corresponding execution times are shown in Fig. 7.3. However, the results contradict our prediction that the processing time should reduce as the number of executors promotes. On the contrary, the execution time decreases at first and then increases after configuring three or more executors, which seems unreasonable. The same phenomenon appears when investigating how the number of cores affects Spark efficiency. Two executors are configured in this test, and the number of cores distributed for the executors are varied from 1 to 4. Similarly, the best result shows up while configuring two cores and two executors. Then the processing time improves with more distributed cores, illustrated in Fig. 7.4. Results from two tests both show inconformity with a linear trend.

A possible explanation is speculated as configured "over- parallelism" issue. It is suggested by Lumen Learning [16] that tunning partition should be reasonable as too many partitions lead to excessive overhead in managing many small tasks. As informed by the official documentation that for local mode, the number of cores on the local machine is equivalent to the default number of parallelism [28]. The

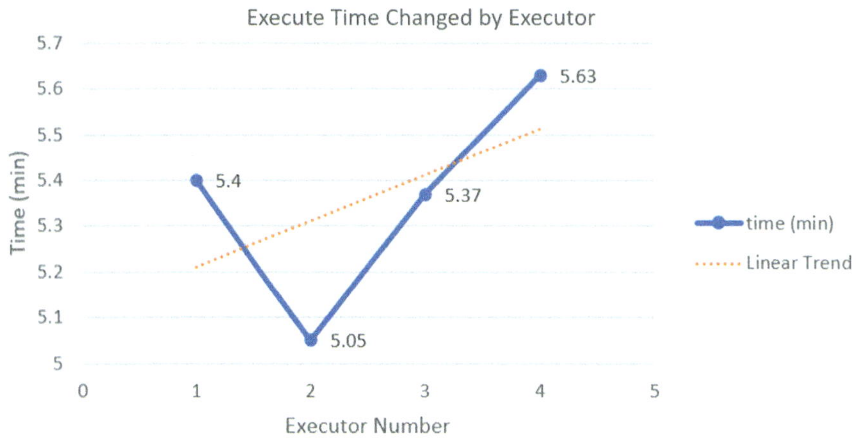

Fig. 7.3 This experiment measures how the execution time of the driver program changed by different numbers of executors. The number of executors increases from 1 to 4. The blue line illustrates the experiment results in a unit of minute, and the orange line demonstrates a linear trend of the experiment results

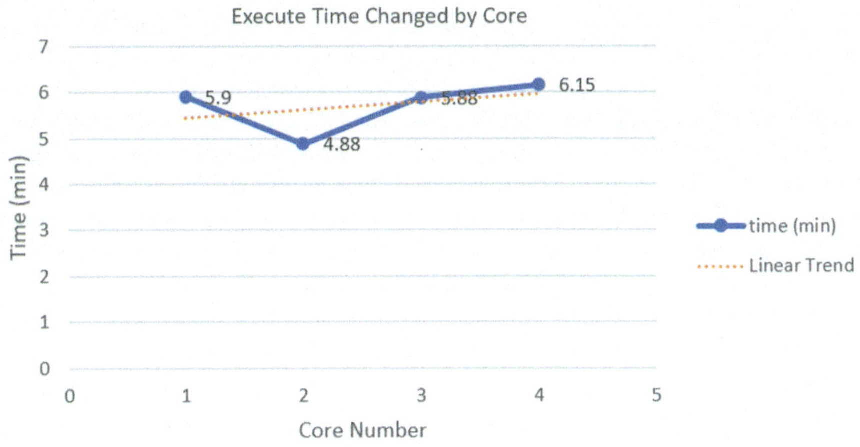

Fig. 7.4 This experiment measures how the driver program's execution time changed by different cores. The number of cores promotes from 1 to 4. The blue line illustrates the experiment results in a unit of minute, and the orange line demonstrates a linear trend of the experiment results

number of tasks that could be processed at once is the total number of cores that could be utilized in the Spark application. If the configured task number is more than the actual amount of tasks that could be processed, other tasks have to wait until a core is available. For our local machine, there are four physical cores available. When the configured degree of parallelism, the number of executors times the number of cores, more significant than the actual number of parallel tasks beyond the partition number, must wait. Therefore, those tests configuring executors or cores more than the actual parallel degree would take more processing time (Figs. 7.3 and 7.4). We can learn from this that in order to execute the driver program more efficiently, the setting partition number should be close to the actual parallel number.

7.5 Conclusions

In this study, we used an effective feature reduction strategy to reduce 433 features to 50 features. This strategy could significantly reduce the running time in model training and testing while the accuracy is not compromised. After testing four models, it was concluded that MLP, compared to SVM, logistic regression, and random forest, has the best credit card fraud detection performance with the IEEE- CIS dataset. Also, our implementation using Spark has good scalability because the increase in the dataset size did not bring the same degree of increase in running time. Thus, it is reasonable to assume that our implementation can handle real-world fraud detection problems with larger datasets. In the process of using Spark, keeping the balance of the Spark partition is very important. Lack of partitions can cause the waste of available cores, and too many partitions lead to the excessive overhead problem. Both can cause a substantial increase in running time. The best parallelization strategy for this implementation is studied and compared. For future work, we plan to use other algorithms, such as Xgboost and Lightgbm, to increase the model performance further. Additionally, the use of larger datasets and the creation of new features may improve the accuracy of predictions. Currently, the dataset we used was collected two years ago. Newer credit card fraud detection datasets with data from in the past two years could be used to test our models and update our models based on the testing result to produce a solution more suitable for real-world deployment.

Acknowledgments The research work in this chapter was supported by our research team members: Ruijie Xiong, Xinyu Gao, Hongyan Deng, and Boya Wang.

References

1. Armbrust M, Xin RS, Lian C, Huai Y, Liu D, Bradley JK, Meng X, Kaftan T, Franklin MJ, Ghodsi A, Zaharia M (2015) Spark SQL: relational data processing in spark. In: Proceedings of the 2015 ACM SIGMOD international conference on management of data, pp 1383–1394. https://doi.org/10.1145/2723372.2742797
2. Belouch M, El Hadaj S, Idhammad M (2018) Performance evaluation of intrusion detection based on machine learning using Apache Spark. Procedia Comput Sci 127:1–6. https://doi.org/10.1016/j.procs.2018.01.091
3. Business Insider (2015) Payments companies are trying to fix the massive credit-card fraud problem with these 5 new security protocols, Available from: http://www.businessinsider.com/how-payment-companies-are-trying-to-close-the-massive-hole-in-credit-card-security-2015-3. Accessed: 02 May 2021
4. Business Wire (2015) Global card fraud losses reach $16.31 Billion—will exceed $35 Billion in 2020 according to the Nilson report, 8. Available from: https://www.businesswire.com/news/home/20150804007054/en/Global-Card-Fraud-Losses-Reach-16.31-Billion-%E2%80%94-Will-Exceed-35-Billion-in-2020-According-to-The-Nilson-Report. Accessed: 02 May 2021
5. Chen M, Mao S, Liu Y (2014) Big data: a survey. Mobile Netw Appl 19(2):171–209. https://doi.org/10.1007/s11036-013-0489-0
6. Chen Y, Han X (2021) CatBoost for fraud detection in financial transactions. In: 2021 IEEE international conference on consumer electronics and computer engineering (ICCECE). IEEE, pp 176–179
7. Dean J, Ghemawat S (2008) MapReduce: simplified data processing on large clusters. Commun ACM 51(1):107–113. https://doi.org/10.1145/1327452.1327492
8. Deng W, Huang Z, Zhang J, Xu J (2021) A data mining based system for transaction fraud detection. In: 2021 IEEE international conference on consumer electronics and computer engineering (ICCECE). IEEE, pp 542–545. https://doi.org/10.1109/ICCECE51280.2021.9342376
9. Excell D (2012) Bayesian inference–the future of online fraud protection. Comput Fraud Secur 2012(2):8–11
10. Ge D, Gu J, Chang S, Cai J (2020) Credit card fraud detection using lightgbm model. In: 2020 international conference on E-commerce and internet technology (ECIT). IEEE, pp 232–236
11. Gopalani S, Arora R (2015) Comparing apache spark and map reduce with performance analysis using k-means. Int J Comput Appl 113(1). https://doi.org/10.5120/19788-0531
12. Illowsky B (2018) Testing the significance of the correlation coefficient. Adapted By darlene young introductory statistics. [Online]. Available at: https://courses.lumenlearning.com/introstats1/chapter/testing-the-significance-of-the-correlation-coefficient/. Accessed: 04 May 2021

13. Jolliffe IT, Cadima J (2016) Principal component analysis: a review and recent developments. Philosophical Trans Royal Soc Math Phys Eng Sci 374(2065):20150202. https://doi.org/10.1098/rsta.2015.0202. Accessed: 26 April 2021
14. Kaggle (2021) IEEE-CIS fraud detection can you detect fraud from customer transactions? 2 May 2021. Available at: https://www.kaggle.com/c/ieee-fraud-detection/data
15. Lippmann R (1987) An introduction to computing with neural nets. IEEE ASSP Mag 4(2):4–22. https://doi.org/10.1109/MASSP.1987.1165576
16. Lumen Learning (n.d.) Testing the significance of the correlation coefficient—introduction to statistics, courses.lumenlearning.com. [Online]. Available at: https://courses.lumenlearning.com/introstats1/chapter/testing-the-significance-of-the-correlation-coefficient/. Accessed: 04 May 2021
17. Madden S (2012) From databases to big data. IEEE Internet Comput 16(3):4–6. https://doi.org/10.1109/mic.2012.50
18. Meng X, Bradley J, Yavuz B, Sparks E, Venkataraman S, Liu D, Freeman J, Tsai DB, Amde M, Owen S, Xin D (2016) Mllib: machine learning in Apache spark. J Mach Learn Res 17(1):1235–1241
19. Mostafaeipour A, Jahangard Rafsanjani A, Ahmadi M, Arockia Dhanraj J (2021) Investigating the performance of Hadoop and Spark platforms on machine learning algorithms. J Supercomput 77(2):1273–1300. https://doi.org/10.1007/s11227-020-03328-5
20. Najadat H, Altiti O, Aqouleh AA, Younes M (2020) Credit card fraud detection based on machine and deep learning. In: 2020 11th international conference on information and communication systems (ICICS). IEEE, pp 204–208. https://doi.org/10.1109/ICICS49469.2020.239524
21. Nodarakis N, Sioutas S, Tsakalidis AK, Tzimas G (2016) Large scale sentiment analysis on twitter with spark. In: EDBT/ICDT Workshops, pp 1–8
22. Panigrahi S, Lenka RK, Stitipragyan A (2016) A hybrid distributed collaborative filtering recommender engine using Apache spark. Procedia Comput Sci 83:1000–1006. https://doi.org/10.1016/j.procs.2016.04.214
23. Roy A, Sun J, Mahoney R, Alonzi L, Adams S, Beling P (2018). Deep learning detecting fraud in credit card transactions. In: 2018 systems and information engineering design symposium (SIEDS). IEEE, pp 129–134
24. Sagiroglu S, Sinanc D (2013) Big data: a review. In: 2013 international conference on collaboration technologies and systems (CTS). IEEE, pp 42–47. https://doi.org/10.1109/CTS.2013.6567202
25. Salloum S, Dautov R, Chen X, Peng PX, Huang JZ (2016) Big data analytics on Apache Spark. Int J Data Sci Anal 1(3):145–164. https://doi.org/10.1007/s41060-0160027-9
26. Shi J, Qiu Y, Minhas UF, Jiao L, Wang C, Reinwald B, Özcan F (2015) Clash of the titans: Mapreduce vs. spark for large scale data analytics. Proc VLDB Endow 8(13):2110–2121https://doi.org/10.14778/2831360.2831365
27. Spark (n.d. a) Hardware provisioning—Spark 3.1.1 documentation. Spark.apache.org. [Online]. Available: https://spark.apache.org/docs/3.1.1/hardware-provisioning.html$#$cpu-cores. Accessed: 03 May 2021
28. Spark (n.d. b) Configuration-spark 3.1.1 documentation, spark.apache.org. [Online]. Available: https://spark.apache.org/docs/3.1.1/configuration.html$\#$memory-management. Accessed: 04 May 2021.
29. Thusoo A, Sarma JS, Jain N, Shao Z, Chakka P, Anthony S, Liu H, Wyckoff P, Murthy R (2009) Hive: a warehousing solution over a map-reduce framework. Proc VLDB Endow 2(2):1626–1629. https://doi.org/10.14778/1687553.1687609
30. WorldPay (2015) Global payments report preview: your definitive guide to the world of online payments. 27 pages report, Available from: https://docplayer.net/9757847-Global-payments-report-preview.html. Accessed: 18 June 2021
31. Yakovlev Y (2021) Very short summary, on 2 May 2021. Available at: https://www.kaggle.com/c/ieee-fraud-detection/discussion/111257

32. Yee OS, Sagadevan S, Malim NHAH (2018) Credit card fraud detection using machine learning as data mining technique. J Telecommun Electron Comput Eng (JTEC) 10(1–4):23–27
33. Yu X, Li X, Dong Y, Zheng R (2020) A deep neural network algorithm for detecting credit card fraud. In: 2020 international conference on big data, artificial intelligence and internet of things engineering (ICBAIE). IEEE, pp 181–183
34. Zaharia M, Xin RS, Wendell P, Das T, Armbrust M, Dave A, Meng X, Rosen J, Venkataraman S, Franklin MJ, Ghodsi A (2016) Apache spark: a unified engine for big data processing. Commun ACM 59(11):56–65. https://doi.org/10.1145/2934664
35. Zareapoor M, Shamsolmoali P (2015) Application of credit card fraud detection: based on bagging ensemble classifier. Procedia Comput Sci 48(2015):679–685
36. Zhang Y, Tong J, Wang Z, Gao F (2020) Customer transaction fraud detection using xgboost model. In: 2020 international conference on computer engineering and application (ICCEA). IEEE, pp 554–558
37. Zojaji Z, Atani RE, Monadjemi AH (2016) A survey of credit card fraud detection techniques: data and technique oriented perspective. arXiv preprint arXiv:1611.06439

Chapter 8
Moving Forward with Big Data Analytics and Smartness

8.1 A Brief Reflection on Big Data Analytics and Smart Urban Systems

By questioning 'what smartness does in the smart city', Baykurt and Raetzsch [11] provide suggestions for analytical approaches to investigate the complexity of urban systems, such as relationships and networks in the built and natural environments. Many scholars refer to smart cities as platforms where living labs and smart transitions could occur [1, 2, 5, 8, 10, 12, 19, 32]. In a way, they provide us with potential urban experimentation with possible institutional transformations or transitions [32]. What usually occurs is through data collection and analysis centers,hence, by considering the city as a data machine [10], we can simply recognise the opportunities for local governance and various management of urban systems through the use of big data. Since the rise of smart devices in the 2000's, we could see growing popularity in developing online or web data platforms [25] that are accessible for various uses. The same patterns are seen with the growth of ICT-based and ICT-driven platforms [15, 16], where digital data flow is immense [18]. Thus, we see no stop in using big data to optimise various systems. The so-called data revolution [28] would mean further consequences for collecting, analysing, and managing data. As seen in top-down contexts, this trend has already become a worry, but in more democratic contexts, slightly better platforms exist that may be more humane and for the benefit of the majority. Nonetheless, we reserve our doubts in both regards.

The nexus between 'big data analytics' and 'smart urban systems' is growing faster than ever. This is mainly due to the rapid growth of data generation, collection, and use. As Cesario [13] puts it well, "*urban environments continuously generate larger and larger volumes of data, whose analysis can provide descriptive and predictive*

S. Pourroostaei Ardakani and A. Cheshmehzangi, *Big Data Analytics for Smart Urban Systems*, Urban Sustainability, https://doi.org/10.1007/978-981-99-5543-5_8

models as valuable support to inspire and develop data-driven Smart City applications. To this aim, Big data analysis and machine learning algorithms can play a fundamental role in bringing improvements in city policies and urban issues". Thus, there is no clear end to this continuous generation and various applications. Nowadays, our access to real-world cases and real-time data helps us understand how big data analytics could respond to—if not resolve—urban challenges. In more popular areas, such as smart urban metabolism, we see prominent attention is given to urban management scenarios, such as those that include "*modern-day technologies dealing with the complex challenges of growing smart cities*" [34]. In this regard, we see the eventual convergence of smart cities with the Internet of Things (IoT) and Big Data (Singh 2022) is almost inevitable:

> The fast growth in the population density in urban areas demands more facilities and resources. To meet the needs of city development, the use of Internet of Things (IoT) devices and the smart systems is the very quick and valuable source. However, thousands of IoT devices are interconnecting and communicating with each other over the Internet results in generating a huge amount of data, termed as Big Data (Singh 2022).

A conference in 2020 held in Shanghai covered a range of big data analytics for cyber-physical systems in smart cities [7]. At this conference, a range of new big applications in smart urban systems were introduced and explored. Before then, we see that research on big data was primarily correlated with the evolution of IoT technologies [37], showing how various tools and technologies help shape "*the state-of-the-art communication technologies and smart-based applications used within the context of smart cities*". Thus, big data could offer many valuable insights into the city and smart urban systems, particularly since it could cross over various sources and data collection platforms. As Chen et al. [14] highlighted, such data's characteristics often refer to unstructured features, which indicate a broad coverage of data collection [9], smart technologies, smart applications, and information exchanges. Nonetheless, ongoing debates remain in regards to planning the smart city, questioning whether smart urban systems could create smart cities, communities, and people; and if so, how these transformations are provided and how data (or big data) is used in genuine favour of societal enhancements. Furthermore, we have already covered some of these debates in other urban-related research, such as sustainable urbanism directions [17] and data-driven approach to analyse changes, trends, and patterns [6]. As mentioned in the introduction chapter, we will continue to follow up on this book's findings and contributions in a forthcoming volume, where we focus more on sector-based contributions, namely smart transport and healthcare systems.

Fig. 8.1 A brief summary of big data glossary in this book in four key areas/clusters of big data characteristics, big data methods, tools and applications, and data-related knowledge

In this book, we have covered a large range of aspects related to 'big data' and 'big data analytics' research and practice.

Our main areas are covered, including (1) big data characteristics, (2) big data methods, (3) tools and applications, and (4) data-related knowledge (such as data science, data storage, data preprocessing, data processing, etc.) (see Fig. 8.1). This comprehensive coverage correlates well with existing literature and research, as shown in Boxes 8.1 to 8.6. As a process in smart urban systems, data search or discovery is important in how data is integrated, modelled, and analysed in a procedural process. In a way, it reflects well on the four stages of data extraction, database creation, data transformation, and loading into a sort of data warehouse for further use and analysis. In addition, as for big data tools, we introduced a range of examples, such as the usual usage of databases, real-time processing (e.g., Spark), measuring systems, data storage and processing, data analysis, and data warehouse. As shown in the case study chapters, for each tool, there is a range of software, methods, formulae, and processes to be taken into consideration.

Box 8.1 Example of ‘Big Data’ Glossary

By: Innovation Collider, Hirani (2016)

Big Data Solutions for Small Firm—by UC Berkley Sutardja Center

“Big Data has been leveraged by a number of large firms to gain a competitive advantage in the retail industry – the likes of Macy’s, Amazon, reveal that big data has been crucial to helping them increase sales by as much as 10%. Kroger CEO David Dilon refers to big data as his ‘secret weapon.’ Some would argue that big data has allowed large organizations to create an oligopoly. Gary Hawkins, in the Harvard Business Review talks of how big data may ‘kill all but the biggest retailers,’ and that access to big data has allowed smaller firms to be relegated to ‘the role of convenience stores”.

Available from: https://scet.berkeley.edu/big-data-solutions-for-small-firms/.

Box 8.2 Example of 'Big Data' Glossary

By: 7wData, 2015

Mini-glossary: Big Data terms you should know

Top 20 terms include:

Analytics	**Data Governance**
Algorithm	**Data Mining**
Behavioural Analytics	**Data Repository**
Big Data	**Data Scientist**
Business Intelligence	**ETL (extract, transform, and load)**
Clickstream Analytics	**Hadoop**
Dashboard	**Legacy system**
Data Aggregation	**Map/reduce**
Data Analyst	**System of record (SOR) data**

Available from: https://7wdata.be/article-data/article-bigdata/mini-glossary-big-data-terms-you-should-know/.

Box 8.3 Example of 'Big Data Technology' Taxonomy

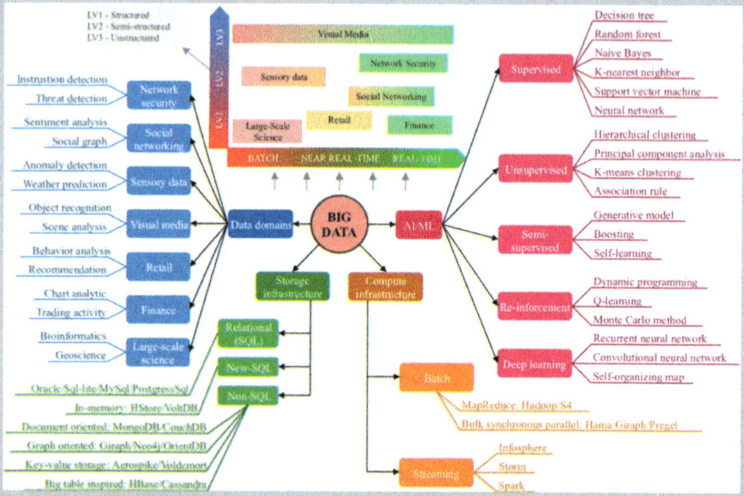

By: Gadekallu et al. (2021)

Taxonomy of big data technology

"Big data has remarkably evolved over the last few years to realize an enormous volume of data generated from newly emerging services and applications and a massive number of Internet-of-Things (IoT) devices. The potential of big data can be realized via analytic and learning techniques, in which the data from various sources is transferred to a central cloud for central storage, processing, and training. However, this conventional approach faces critical issues in terms of data privacy as the data may include sensitive data such as personal information, governments, banking accounts. To overcome this challenge, federated learning (FL) appeared to be a promising learning technique".

Available from: https://creativecommons.org/licenses/by-nc-nd/4.0/.

Box 8.4 Example of 'Big Data' Characteristics

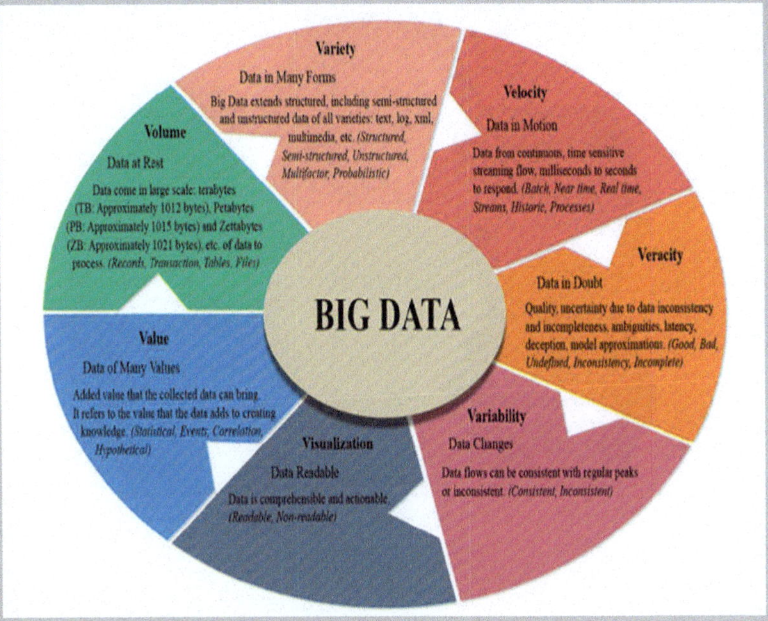

By: Malik et al. (2018)

Big Data Characteristics

> "*Because of the broad utilization of web-based social networking, data is produces by the fast increment. Big Data is giving the office to accumulate, store, oversee and examine information in colossal volume that is produced through the healthcare system. Cloud Computing is an advancement too that insures the fulfillment of IT requirements in a suitable way by providing the cloud-based environment for medical field. Storage is an immense issue for BD, volume of data is huge, this issue may resolve with the help of cloud computing by providing the storage space for data and processing mechanism as well*".

Available from: https://www.matec-conferences.org/articles/matecconf/abs/2018/48/matecconf_meamt2018_03010/matecconf_meamt2018_03010.html.

Box 8.5 Example of 'Big Data' Characteristics

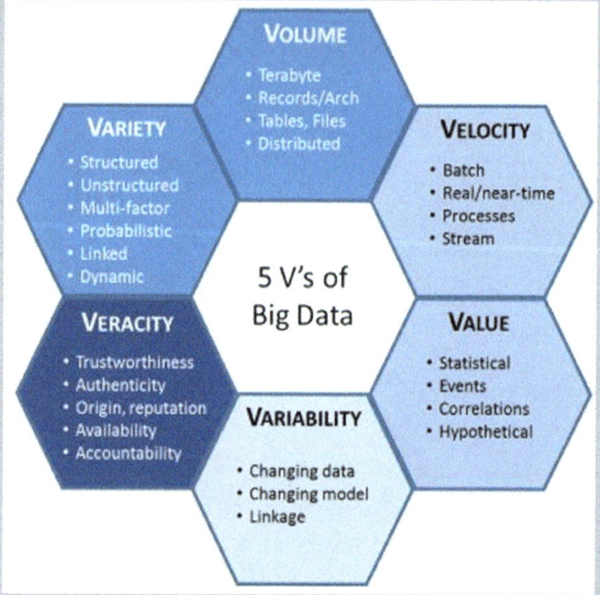

By: Shaqiri (2017)

Big Data Characteristics

"Big data is a concept that involves a combination of technologies and strategies aimed at the collection, analysis, storage, and use f large amount of data. Big data is characterised by data that have a large volume, massive velocity, numerous variety, useful value, and variability. Big data involve various processes that involve the collection, processing, and multi-usage of the collected data. The dynamic nature of big data presents various challenges associated with the data processing and analysis activities. The security and privacy of big data concepts are some of the most pertinent issues associated with big data in the modern world. There have been concerned efforts geared towards the development of robust techniques aimed at solving the various security challenges in big data".

Available from: https://doi.org/10.13140/RG.2.2.23201.10089.

Box 8.6 Example of 'Big Data' and 'Data Science' Glossary

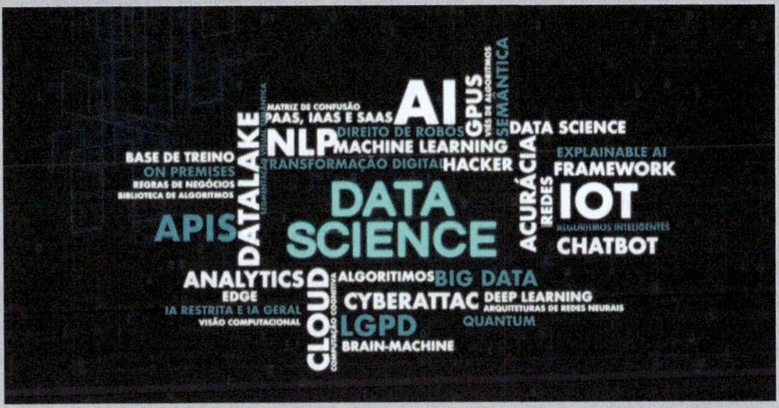

By: i2AI

Comprehensive Glossaries

Available from: https://www.i2ai.org/content/glossary/.

8.2 Methodological Contributions of the Book

Since the 1970s, there is a gradual evolution of analytics terminology across multiple sectors. At first, in the 1970s, it was regarded as 'Decision Support Systems'. In the 1980s, it was developed into 'Enterprise Information Systems', and later into 'Executive Information Systems' in the 1990s. With the emergence of the 'Business Intelligence term in the 2000s, the term is then developed into 'Big Data Analytics' in the 2010s [20]. There is no doubt that it will continue to evolve and become a new terminology in the near future. While the book's case study examples compared some of the traditional and novel methods in research, we also tried to explore some of the common (contemporary) methods in computer science that could be applied to urban-related research, particularly for smart cities and smart urban systems. For us, such applications are important as we expand on how big data analytics has evolved, particularly in cities and the development of smart urban systems. The following is a summary of chapter-by-chapter methodological contributions of the book:

Chapter 2—A time-series big data analysis method is mainly considered using the LSTM model for prediction modelling and analysis. Data pre-processing is followed by pattern selection, correlated with feature selection results and analysis.

Chapter 3—A time-series machine learning approach is mainly used for data pattern recognition, correlated with a Dynamic Time Wrapping (DTW) K-means clustering approach, random forest, and time-series pattern recognition.

Chapter 4—Using two machine learning models, including 'Decision Tree' and 'Gradient Boosting', trained, tested, and evaluated to find the best-fitted techniques. The big data-enabled data preprocessing approach is proposed to prepare the dataset.

Chapter 5—A federated regression analysis is used based on a regression-enabled machine learning model to predict daily travel using the pandemic's situation. It also uses the correlation matrix method to find relevant correlations.

Chapter 6—An adaptive feature section approach (Random Forest) is used in compared to traditional feature selection techniques, such as 'Linear Discriminant Analysis' and 'Principal Component Analysis'.

Chapter 7—A big data classification solution is used based on the availability of index terms, credit fraud detection, and machine learning research. Spark MLlib and Principle Component Analysis (PCA) are used in the chapter.

Lastly, the book also highlights the change in the overall landscape of big data research and big data technologies (Fig. 8.2). It verifies the importance of the big data analytics process, including three overarching stages of data pre-processing, data processing, and data analysis and use (Fig. 8.3). The findings and methodological contributions from this book help future big data analysis research, particularly for urban management and smart urban systems. They could partly support (urban) decision-making processes, smart transitions, and urban governance processes.

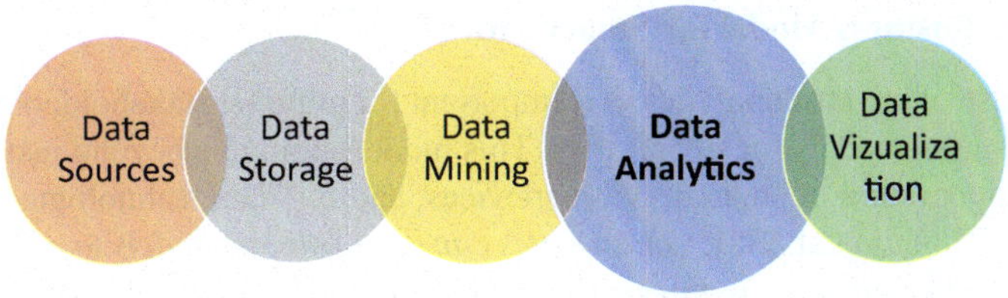

Fig. 8.2 The change in the overall landscape of big data research and big data technologies represents rapid changes that occur in the area of big data analytics, in particular in correlation with available technologies. The suggested big data ecosystem indicates five key areas or tools commonly used in research and are highlighted in this book. The first is 'Data Sources', which could include both internal and external data. The second is 'Data Storage', which commonly refers to big data storage software tools store as well as methods of managing and retrieving big data. The third is 'Data Mining', the set of tools for data analysis, finding relationships, patterns, anomalies, etc. The fourth is 'Data Analytics', which refers to specific and more advanced software specifically with advanced analytical capabilities. And lastly, the fifth is 'Data Visualization', which is based on available software to present data through analytics tools and the development of visuals like graphs, bars, charts, etc. *Source* Redrawn by the authors and inspired from Intellspot (n.d.)

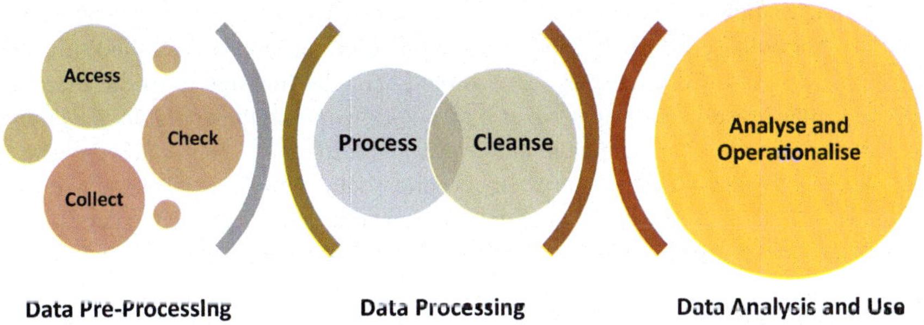

Fig. 8.3 The summary of big data analytics process

8.3 Concluding Remarks: A Summary of Lessons Learnt for Future Research

It is hard to summarise a wide range of contributions in a brief conclusion section. Hence, we intentionally leave this section open as a summary of lessons learnt for future research directions, particularly focused on big data analytics and its applications in smart urban systems. As suggested by Zeng et al. [40], "*Big Data Analytics can be leveraged to support a city's transformation into a smart destination*". Thus, we see transformative opportunities for the future of cities and communities around the globe. Here, we summarise five overarching lessons learnt for future research, particularly from the perspective of big data analytics in smart urban systems.

- **Future Business Model for Smart Cities**

With the use of big data analytics, it is important for multiple stakeholders to develop future business models for smart cities. This model should be beyond just the inclusion and integration of smart services, devices, and big data technologies. As highlighted by Timeus et al. [38], "*smart cities can use business models to evaluate what value they offer citizens by integrating ICT into their infrastructure and services*" (also see: [21, 23, 24, 31, 35]). The business model should help provide opportunities for smart services to have an impact on society, the economy, and the environment to the least. Thus, the efficiency of such a business model is vital in the technology sector, planning, and sustainable urban development pathways. The smart urban systems could then be developed into self-sustaining and multi-modal systems with integrated services and amenities and people-oriented operations.

- **Data-driven Smartness for Cities and Communities**

A data-driven smart city is often regarded as a city with successful (urban) systems enabling the functioning of the data-driven platforms, such as infrastructure and operations of the city for data collection and use purposes. Usually, we refer to such an approach emphasizing smart technologies and performance-based mechanisms, where cities could generate big data and help use such real-time data and information as effectively and efficiently as possible [27]. The findings from the literature illustrate that "*the current new wave of smart cities with real time data are promoting citizen participation focusing on human, social capital as an essential component in future cities*". This is mainly due to their complex social networks and physical infrastructure [22] that suggest facilitating data-driven smart applications and services for cities and communities, urban infrastructure, and urban governance.

- **Data Integration**

One of the current challenges for big data and smart cities is the lack of data integration in urban systems or across multiple systems. As Ribeiro and Braghetto [33] suggest, "*the data generated by smart cities have low integration, as the systems that produce them are usually closed and developed for specific needs. Moreover, the large volume of data, and the semantic and structural changes in datasets over time make the use of data to support decision-making even more difficult*". Thus, from the findings of this book, we also suggest having data integration solutions, particularly to create opportunities to uncover data features and limitations. Examples of prototyping applications and platforms for integrating heterogeneous data create integration opportunities for experimenting with IoT and other data platforms. In this regard, the approach for data fusion in smart city applications [29] is an essential task for future urban development plans and directions. One of the key aspects would be to establish smart city data fusion platforms where data can be integrated and utilised in a smarter (or more innovative) way.

- **Context-specific Solutions**

In our previous work, we also highlighted the importance of context-specific solutions. When it comes to big data analytics, contextual factors are also important as they determine how data is used and managed. There are apparent contextual differences in how data is used in different countries or locations, meaning that there are differences in how data sensitivity, data protection, and data privacy are considered [18]. Thus, smart urban systems need to be fully reviewed, analysed compared, and classified based on contextual conditions. In particular, big data and data mining methods (and technologies) should develop and secure a sort of big data ecosystem architecture [3] that could respond well to local/individualised/contextual challenges and solutions.

- **Cross-Sectoral Approach**

A cross-sectoral or cross-domain approach has become solution to many of the current data usage and analysis problems. The assumption that data should be generated and used in one sector is a wrong perception, meaning that big data tools and technologies could help balance between generation and data analysis beyond just one sector. For the public benefit, in particular, we see a growing demand for cross-sectorial big data use and analysis [30]. As suggested by Yang and Ji [39], "*many wicked problems, such as environmental issues, require organizations from multiple sectors to form cross-sectoral alliances. Cross-sectoral alliance networks can transfer resources and signal affiliations and value alignment between strategic partners. The communication of cross-sectoral alliances is a form of CSR communication that serves organizations' strategic goals and objectives*". In a way, this alliance and partnership between multiple sectors is essential. Thus, the future directions could be moving towards big data accessibility or big data as a self-service platform. Thus, "*The proliferation of Big Data applications puts pressure on improving and optimizing the handling of diverse datasets across different domains*" [4]. In other words, breaking the barriers between sectors and instead establishing and promoting cooperation and partnerships.

Lastly, we conclude the book by reflecting on current research gaps and challenges, some of which are highlighted above in five possible directions. We touch on the importance of having ideas and platforms for 'transformative urban transitions'. The vision of big data analytics to enhance smart urban systems could help us develop a better nexus between smartness and other important factors in cities, such as resilience, healthy living environments, green development, sustainable development, etc. Game-changing transitions are needed to develop smart-resilient, smart-healthy, smart-green, and smart-sustainable cities, rather than just letting technologies dominate us through transitions of becoming different beings. Thus, the role of big data analytics in smart urban systems should develop further to help us utilise smart technologies, ICTs, and data-driven platforms more smartly and efficiently. The move should diverge from the current pure control and management trends and help create better places for people. In doing so, smart urban systems are utilised for better living and working environments, and we will not have the assumption of living in a matrix-type place. Thus, we urge researchers, practitioners, and governments to develop further partnerships to avoid over-reliance on technologies and instead use technologies and big data analytics in our favour, i.e., to make places and people smarter, to make healthier and liveable environments for all, and to create pathways for the future of cities.

References

1. Albino V, Berardi U, Dangelico RM (2015) Smart cities: definitions, dimensions, performance, and initiatives. J Urban Technol 22(1):3–21
2. Almirall E, Lee M, Wareham J (2012) Mapping living labs in the landscape of innovation methodologies. Technol Innov Manage Rev 2:12–18
3. Anwar MJ, Gill AQ, Hussain FK, Imran M (2021) Secure big data ecosystem architecture: challenges and solutions. EURASIP J Wirel Commun Netw. Article Number: 130. https://doi.org/10.1186/s13638-021-01996-2
4. Arapakis I, Becerra Y, Boehm O, Bravos G, Chatzigiannakis V, Cugansco C, Demetriou G, Eleftheriou I, Mascolo JE, Fodor L, Ioannidi S, Jakovetic D, Kallipolitis L, Kavakli E, Kopanaki D, Kourtellis N, Marcos MM, de Pozuelo RM, Milosevic N, Morandi G, Montanera EP, Ristow G, Sakellariou R, Sirvent R, Skrbic S, Spais I, Vasiliadis G, Vinov M (2019) Towards specification of a software architecture for cross-sectoral big data applications. In: 2019 IEEE world congress on services (SERVICES), Conference held in Milan, Italy. https://doi.org/10.1109/SERVICES.2019.00120
5. Araya D (ed) (2015) Smart cities as democratic ecologies. Palgrave Macmillan, Basingstoke
6. Ardakani PS, Xia T, Cheshmehzangi A, Zhang Z (2021) An urban-level prediction of lockdown measures impact on the prevalence of the COVID-19 pandemic. Genus 78:28. Available from https://doi.org/10.1186/s41118-022-00174-6
7. Attiquzzaman M, Yen N, Xu Z (2021) Big data analytics for cyber-physical system in smart city: BDCPS 2020, 28–29 Dec 2020, Shanghai, China. Part of the Advances in Intelligent Systems and computing (AISC), vol 1303. Springer, Singapore
8. Aymen A, Kachouri A, Mahfoudhi A (2017) Data analysis and outlier detection in smart city. In: 2017 international conference on smart, monitored and controlled cities (SM2C). https://doi.org/10.1109/SM2C.2017.8071256. Also available from https://ieeexplore.ieee.org/document/8071256/
9. Batty M (2013) Big data, smart cities and city planning. Dialogues Human Geogr 3(3). https://doi.org/10.1177/2043820613513390
10. Baykurt B (2019) The city as data machine: local governance in the age of big data. Doctoral dissertation, Columbia University
11. Baykurt B, Raetszch C (2020) What smartness does in the smart city: from vision to policy. Convergence Int J Res New Media Technol 26(4). https://doi.org/10.1177/1354856520913405
12. Bifulco F, Tregua M, Amitrano CC (2017) Co-governing smart cities through living labs. Top evidences from EU. Transylvanian Rev Adm Sci 50:21–37
13. Cesario E (2023) Big data analytics and smart cities: applications, challenges, and opportunities. Frontiers Big Data 6. Section on Data Mining and Management. https://doi.org/10.3389/fdata.2023.1149402
14. Chen M, Mao S, Liu Y (2014) Big data: a survey. Mob Netw Appl 19:171–209. https://doi.org/10.1007/s11036-013-0489-0
15. Cheshmehzangi A (2022) ICT, cities, and reaching positive peace. Springer, Singapore
16. Cheshmehzangi A (2022b) The application of ICT and smart technologies in cities and communities: an overview. In: ICT, cities, and reaching positive peace. Springer, Singapore, pp 1–16

17. Cheshmehzangi A, Dawodu A, Sharifi A (2021) Sustainable urbanism in China. Routledge, New York
18. Cheshmehzangi A, Su Z, Zou T (2023) ICT applications and the COVID-19 pandemic: impacts on the individual's digital data, digital privacy, and data protection. Frontiers Human Dyn 5. Section on Digital Impacts. Available from https://doi.org/10.3389/fhumd.2023.971504
19. Davenport TH (2017) How analytics has changed in the last 10 years (and how it's stayed the same). Harvard business review: analytics services. Available from https://hbr.org/2017/06/how-analytics-has-changed-in-the-last-10-years-and-how-its-stayed-the-same
20. Delen D, Ram S (2018) Research challenges and opportunities in business analytics. J Bus Anal 1(1):2–12. https://doi.org/10.1080/2573234X.2018.1507324. Also available from https://www.tandfonline.com/doi/full/https://doi.org/10.1080/2573234X.2018.1507324
21. Díaz-Díaz R, Muñoz L, Pérez-González D (2017) The business model evaluation tool for smart cities: application to smartsantander use cases. Energies 10(3):262. https://doi.org/10.3390/en10030262
22. Foth M, Choi JHJ, Satchell C (2011) Urban informatics. In: Proceedings of the ACM 2011 conference on computer supported cooperative work, Hangzhou, China, 19–23 Mar 2011, pp 1–8
23. Gascó M (2016) What makes a city smart? Lessons from Barcelona. In: Presented at the Hawaii international conference on system science, Kauau, Hawaii, USA
24. Grossi G, Reichard C (2008) Municipal corporatization in Germany and Italy. Public Manage Rev 10(5):597–617. https://doi.org/10.1080/14719030802264275
25. Helmond A (2015) The platformization of the web: making web data platform ready. Soc Media Soc 1(2):1–11
26. Intellspot (n.d.) Big data technologies: list, stack, and ecosystem in demand, part of business intelligence. Available from https://www.intellspot.com/big-data-technologies/?ref=quuu
27. Kaluarachchi Y (2022) Implementing data-driven smart city applications for future cities. Smart Cities 5(2):455–474. https://doi.org/10.3390/smartcities5020025
28. Kitchin R (2014) The data revolution: big data, open data, data infrastructures and their consequences. Sage Publications, Thousand Oaks
29. Lau BPL, Marakkalage SH, Zhou Y, Ul Hassan N, Yuen C, Zhang M, Tan U-X (2019) A survey of data fusion in smart city applications. Inform Fusion 52:357–374. https://doi.org/10.1016/j.inffus.2019.05.004
30. Laurie GT (2019) Cross-sectoral big data. ABR 11:327–339. https://doi.org/10.1007/s41649-019-00093-3
31. Paskaleva KA (2009) Enabling the smart city: the progress of city E-governance in Europe. Int J Innov Reg Dev 1(4):405–422. https://doi.org/10.1504/IJIRD.2009.02273
32. Raven R, Sengers F, Spaeth P, Xie L, Cheshmehzangi A, de Jong M (2017) Urban experimentation and institutional arrangements, Eur Plann Stud 27(2). Urban Experimentation and Sustainability Transitions. https://doi.org/10.1080/09654313.2017.1393047
33. Ribeiro MB, Braghetto KR (2021) A data integration architecture for smart cities. In: Conference: Simpósio Brasileiro de Banco de Dados (SBBD), Proceedings of the 36th Brazilian symposium on databases. https://doi.org/10.5753/sbbd.2021.17878. Available from https://sol.sbc.org.br/index.php/sbbd/article/view/17878
34. Ruchira G, Sengupta D (2023) Smart urban metabolism: a big-data and machine learning perspective. In: Bhadouria R, Tripathi S, Singh P, Joshi PK, Singh R (eds) Urban metabolism and climate change: perspectives for sustainable cities. Springer, Cham, pp 325–344. https://doi.org/10.1007/978-3-031-29422-8_16
35. Schaffers H, Komninos N, Pallot M, Trousse B, Nilsson M, Oliveira A (2011) Smart cities and the future internet: towards cooperation frameworks for open innovation. In: Domingue J, Galis A, Gavras A, Zahariadis T, Lambert D, Cleary F, Nilsson M (eds) The future internet, vol 6656. Springer, Berlin, Heidelberg, pp 431–446. http://link.springer.com/https://doi.org/10.1007/978-3-642-20898-0_31
36. Singh A, Kumar A (2022) Convergence of smart city with IoT and big data. Available from https://sciforum.net/paper/view/13930

37. Targio Hashem IA, Chang V, Anuar NB, Adewole K, Yaqoob I, Gani A, Ahmed E, Chiroma H (2016) The role of big data in smart city. Int J Inform Manage 36(5):748–758. https://doi.org/10.1016/j.ijinfomgt.2016.05.002
38. Timeus K, Vinaixa J, Pardo-Bosch F (2020) Creating business models for smart cities: a practical framework. Public Manage Rev 22(5): Special issue: Management, Governance and Accountability for Smart Cities and Communities.https://doi.org/10.1080/14719037.2020.1718187
39. Yang A, Ji YG (2019) The quest for legitimacy and the communication of strategic cross-sectoral partnership on Facebook: a big data study. Public Relat Rev 45(5):101839. https://doi.org/10.1016/j.pubrev.2019.101839
40. Zeng D, Tim Y, Yu J, Liu W (2020) Actualizing big data analytics for smart cities: a cascading affordance study. Int J Inform Manage 54:102156. https://doi.org/10.1016/j.ijinfomgt.2020.102156

Index

S. Pourroostaei Ardakani and A. Cheshmehzangi, *Big Data Analytics for Smart Urban Systems*, Urban Sustainability, https://doi.org/10.1007/978-981-99-5543-5